NATO ASI Series
Advanced Science Institutes Series

A series presenting the results of activities sponsored by the NATO Science Committee, which aims at the dissemination of advanced scientific and technological knowledge, with a view to strengthening links between scientific communities.

The Series is published by an international board of publishers in conjunction with the NATO Scientific Affairs Division

A	Life Sciences	Plenum Publishing Corporation
B	Physics	London and New York
C	Mathematical and Physical Sciences	Kluwer Academic Publishers
D	Behavioural and Social Sciences	Dordrecht, Boston and London
E	Applied Sciences	
F	Computer and Systems Sciences	Springer-Verlag
G	Ecological Sciences	Berlin Heidelberg New York Barcelona
H	Cell Biology	Budapest Hong Kong London Milan
I	Global Environmental Change	Paris Santa Clara Singapore Tokyo

PARTNERSHIP SUB-SERIES

1. Disarmament Technologies	Kluwer Academic Publishers
2. Environment	Springer-Verlag
3. High Technology	Kluwer Academic Publishers
4. Science and Technology Policy	Kluwer Academic Publishers
5. Computer Networking	Kluwer Academic Publishers

The Partnership Sub-Series incorporates activities undertaken in collaboration with NATO's Cooperation Partners, the countries of the CIS and Central and Eastern Europe, in Priority Areas of concern to those countries.

NATO-PCO DATABASE

The electronic index to the NATO ASI Series provides full bibliographical references (with keywords and/or abstracts) to about 50000 contributions from international scientists published in all sections of the NATO ASI Series. Access to the NATO-PCO DATABASE compiled by the NATO Publication Coordination Office is possible in two ways:

- via online FILE 128 (NATO-PCO DATABASE) hosted by ESRIN,
  Via Galileo Galilei, I-00044 Frascati, Italy.

- via CD-ROM "NATO Science & Technology Disk" with user-friendly retrieval software in English, French and German (© WTV GmbH and DATAWARE Technologies Inc. 1992).

The CD-ROM can be ordered through any member of the Board of Publishers or through NATO-PCO, Overijse, Belgium.

Series F: Computer and Systems Sciences, Vol. 145

The NATO ASI Series F Special Programme on
ADVANCED EDUCATIONAL TECHNOLOGY

This book contains the proceedings of a NATO Advanced Research
Workshop held within the activities of the NATO Special Programme on
Advanced Educational Technology, running from 1988 to 1993 under the
auspices of the NATO Science Committee. The books published so far in
the Special Programme are listed briefly, as well as in detail together with
other volumes in NATO ASI Series F, at the end of this volume.

Springer

Berlin
Heidelberg
New York
Barcelona
Budapest
Hong Kong
London
Milan
Paris
Santa Clara
Singapore
Tokyo

Advanced Educational Technology: Research Issues and Future Potential

Edited by

Thomas T. Liao

Department of Technology and Society
College of Engineering and Applied Sciences
State University of New York at Stony Brook
Stony Brook, New York 11794-2250, USA

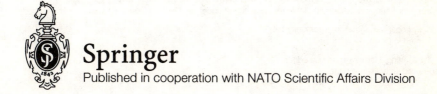

Springer
Published in cooperation with NATO Scientific Affairs Division

Proceedings of the NATO Advanced Research Workshop on Advanced Educational Technology – Research Issues and Future Potential, held in Grenoble, France, September 25–28, 1993

Library of Congress Cataloging-in-Publication Data

Advanced educational technology : research issues and future potential
/ edited by Thomas T. Liao.
        p.   cm. -- (NATO ASI series. Series F, Computer and systems
sciences ; vol. 145)
    "Published in cooperation with NATO Scientific Affairs Division."
    "Proceedings of the NATO Advanced Research Workshop on Advanced
Educational Technology--Research Issues and Future Potential, held
in Grenoble France, September 25-28, 1993"--T.p. verso.
    Includes bibliographical references and index.
    ISBN 3-540-59090-0 (alk. paper)
    1. Educational technology--Congresses. 2. Educational technology-
-Research--Congresses. 3. Computer-assisted instruction-
-Congresses. 4. Intelligent tutoring systems--Congresses.
5. Information technology--Congresses.   I. Liao, Thomas T.
II. NATO Advanced Research Workshop on Advanced Educational
Technology--Research Issues and Future Potential (1993 : Grenoble,
France) III. Series: NATO ASI series. Series F, Computer and
systems sciences ; no. 145.
LB1028.3.A35   1996
371.3'078--dc20   95-50355
                                                          CIP

CR Subject Classification (1991): K.3, H.4, I.2, J.2

ISBN 3-540-59090-0 Springer-Verlag Berlin Heidelberg New York

© Springer-Verlag Berlin Heidelberg 1996
Printed in Germany

Typesetting: Camera-ready by editor
Printed on acid-free paper
SPIN: 10486101   45/3142 – 5 4 3 2 1 0

Preface

As we approach the 21st century, the need to better link research findings and practical applications of advanced educational technologies (AET) continues to be a priority. During the five-year NATO Special Programme on AET, many advanced study institutes and research workshops focused on building bridges between researchers in and users of educational technology. The organizing committee of the final capstone workshop which took place in September 1993 also chose to focus on this theme.

Three position papers, written by members of the AET advisory committee, provided the background and platform for the two-day workshop that was designed to provide guidelines for future AET research and implementation projects. Nicolas Balacheff kicked off the workshop with a philosophical review of the research issues and future research agendas. Herman Bouma and his colleagues at the Institute for Perception Research discussed implementation issues and problems of technology transfer from research laboratories to educational product development. These two position papers provided the foundation for papers that focused on the following AET topics:

- Interactive Systems
- Contact-Based Applications
- Instructional Design

Workshop participants had ample opportunity to react and offer suggestions for improving the papers. Philippe Duchastel concluded the conference with a provocative description of his vision of the future of AET. His view of instruction and learning is that:

> "Learning interfaces are the windows on the world through which a person views information and which cause a certain quality of learning to occur."

Using advanced educational technology systems provides learners with windows to the world that have the potential of helping people make better sense of the world. The challenge to AET Research & Development professionals is to perfect prototype learning environments and make them useful and accessible to all learners.

The NATO Special Programme on AET (1987–1994), besides providing an opportunity for hundreds of professionals to share their expertise has also resulted in about thirty books on various AET topics. Hopefully, this documentation of the "best" practice in research and implementation will provide inspiration and ideas to guide future AET activities. The management of the AET Special Programme was provided by Luis V. da Cunha and his seven-year contribution to the success of the programme is much appreciated. I also want to acknowledge the help of Romayne Dickinson and Caroline dos Santos for providing the administrative support for the capstone workshop. Editorial assistance for this book was provided by Joanne English Daly.

It is my hope and that of the members of the advisory and organizing committees that this collection of papers from the capstone workshop will be useful as tools to inform and guide future AET activities. It is the beginning of greater international collaboration and cooperation.

SUNY at Stony Brook   Thomas T. Liao
October 1995

Table of Contents

1

Advanced Educational Technology: Knowledge Revisited

N. Balacheff

DidaTech, Laboratoire LSD2, IMAG-CNRS & Université Joseph Fourier
BP 53 38041 Grenoble Cedex 9, France
E-mail: Nicolas.Balacheff@imag.fr

Abstract: AET R&D cannot avoid the question of the nature of knowledge which is at the core of both learning and teaching or training. The way this problem can be handled for the purpose of design and implementation of systems supporting human learning, the question of knowledge representations for the purpose of computational models as well as the question of the place of knowledge in person/machine interactions suggest that knowledge should be revisited in the light of the AET research programme. In this chapter I consider this question *from* the point of view of computational modeling and situated AET.

Keywords: Epistemology, artificial intelligence, computational transposition, knowledge modeling, learner modeling, human-machine interaction, didactical contract, intelligent learning environment, educational technology.

1 Introduction

The scope statement of the NATO Special Programme on Advanced Educational Technology (AET), published in 1988, refers to the "applications of new technologies for delivering *instruction*, such as the microcomputer, interactive video disk systems and electronic communication". This statement adds "the detailed design of the instructional material is based on models of learning and teaching developed from the contemporary literature of cognitive science and related areas". Indeed, this domain of R&D involves the collaboration of researchers from several different disciplines, from sciences for computer engineering to human and social sciences. The aim of the program was to facilitate their interaction and collaboration. This statement suggests that designers of AET systems could borrow theories and models from "cognitive sciences and other areas" in order to implement them. The dual expectation could have been for researchers in "cognitive sciences and other areas" to borrow tools and methods from Technology R&D in order to implement their own models. But both expectations were deceived. It is now clear that there is no ready-made educational theories and models which could be straight forwardly implemented, and no technological tools available for the mere implementation of a well designed educational material. As emphasized by Winne and Jones in the scientific report of their ASI[1] "research and development in instructional computing systems have not strongly linked two fields that are foundational in this work – computing-science and instructional science". Actually, they could have mentioned other disciplines and research areas, like psychology, cognitive ergonomy, HCI, artificial intelligence, telecommunication, as well as domains of reference such as physics, language, mathematics, medicine, etc.

A first and essential lesson from this NATO special programme is the need for an integrated collaboration between these disciplines, which implies the capacity to develop shared concepts, theoretical frameworks and methodologies.

The keystone of this collaboration is *knowledge*. No research program in the field of AET R&D can avoid the question of the nature of knowledge which is at the core of both learning and teaching or training. The way it can be handled for the purpose of the design and the implementation of systems supporting human learning, the question of its representation for the purpose of computational models is present in all the discussions, papers and reports produced during this program. But how can we develop a shared view of knowledge coping with computational constraints, psychological theories of learning, educational theories or principles of person/machine interaction? This is the central question we have to consider with priority in the coming decade. Knowledge should be revisited in the light of the AET research programme.

A starting point could be the question of the evaluation of AET systems. It usually comes as the last chapter of books or articles. In the scope statement of the programme, it came in the last section; its priority was supported by the following remark "the best designed and developed instructional systems if poorly implemented will fail". In my opinion, this statement bears witness to a common confusion between the technical quality and robustness of AET and its educational adequacy. Actually the notion of "failure" of an AET system has more or less been left to intuition, it might be time to question it and to try to clarify it as much as possible. Since the fundamental aim of AET systems is to support learning, and since learning can only be understood with reference to some content, it is from this point that we can try to identify the main research issues in our domain.

The second lesson of this programme may be that research on Advanced Educational Technology cannot avoid epistemology. AET is a very large domain, it is not in my capacity to consider all of its aspects. One could identify two large

---

[1] "Synthese of instructional and computing science for effective instructional computing systems", Calgary, Canada, July 15–27 1990. Directors: Dr. Philip H. Winne, Dr. Marlene Jones.

categories of technology-based environments: technologies which support human/ human interactions, for example allowing telepresence, and technologies which in some sense tend to "replace" human beings, for example tutoring systems. Indeed the dividing line is not rigid, but it might help to accept this dichotomy. In particular a convergence is obvious around the question of development of groupware, and of various support for collaboration – for example, users being distant but sharing the same environment and communicating through a video channel (Smith 1988).

In the present chapter, I will focus on AET which aims at allowing some learning in an essentially autonomous way. To prepare the paper, since I aim to make not a survey but a contribution to the summary of the NATO Special Programme on Advanced Educational Technology, I have mainly used as a bibliographical source the volumes published in NATO ASI Series F (Springer-Verlag) in the context of this programme and some papers published in the Journal of Artificial Intelligence in Education.

2   Epistemological Issues

2.1 Knowledge: Stake of the Interaction

The explicit aim of AET is to allow its user to learn some knowledge as the result of her/his interaction with a physical device. Whether this interaction addresses this aim directly, like in the course of a dialogue in the so-called "front teaching", or indirectly, like in "learning by discovery" strategies, in all the cases the nature of the interaction, the way it is managed, binds the knowledge constructed by the learner. Many questions are related to the management of interaction, I will consider some of them in Section 3.2. But answering these questions means that one has clarified what is meant by "knowledge" which is the real stake of this interaction.

A classical distinction between "declarative knowledge" and "procedural knowledge" has for a long time influenced the design of AI-based AET. The former gives ground to environments based on dialogic interaction, the later gives ground to environments more centered on the acquisition of problem solving skills. The paradigm of procedural knowledge has been reinforced by the efficiency of rule-based systems and the initial *problématique* of AET as a question of transfer of problem-solving expertise and skills. Ohlsson (1988) discusses this distinction which he relates to the more basic dichotomy between "cognitive skills" and "abstract knowledge". Following his distinction, one can have a cognitive skill, that is the capacity *to do,* without the knowledge, or conversely one could have some knowledge but not the related skill. This distinction leads Ohlsson to state that "it is the ideas that constitute the core of the curriculum, not their applications" (ibid., p. 77) – which might be dangerous. Whatever the debate provoked by Ohlsson's position, it raises clearly the complexity of capturing the nature of knowledge as the object of an educational transaction – it is interesting that this author views the problem differently when he refers to "academic schooling" and when he refers to vocational training. The a-contextual nature assigned to abstract knowledge in this approach is so radical that the question of its validity is considered as not being relevant for the on going discussion (ibid., p. 83). Self (1992) goes even further suggesting the substitution of "beliefs" to "knowledge" in the philosophy of Intelligent Tutoring Systems (ITS). This author invites us to consider even the knowledge to be taught as belief. The advantage of Self position is to facilitate a transfer of some AI tools and concepts (belief's logic, belief revision) to ITS research, but it has the strong disadvantage of creating a confusion between two major epistemological concepts: beliefs and knowledge (Popper 1979, p. 3). A fallibilist epistemology of knowledge must not lead to confuse it with beliefs, the very meaning of fallibilism is that knowledge "can" be refuted and that its growth is the result of a dialectic of proofs and refutations. Beliefs are basically out of the scope of any refutation. The recent emergence of situated cognition opens a new theoretical perspective on knowledge, giving to context a place "lost" in the search for the essence of knowing. But sometimes, it carries a radical view of knowledge which is always re-created in situations (Clancey 1993, 14-15).

All of these positions, and others I do not mention here because of the lack of space, are at the core of a general debate about the nature of knowledge and its implication for AET research and development. I would very much like to contribute to this debate, but it is not the purpose of the present paper, so I will just tentatively summarize points which constitute a common ground:

- knowledge cannot be reduced to a text;
- students don't receive passively knowledge, they are active constructors of meaning;
- errors are symptoms of the nature of learners' conceptions;
- learning is a dynamic process.

These statements, but in different ways and with different consequences, are now largely shared.

Two more principles (Vergnaud 1992) must be added as starting points for theoretical foundations of our domain: (i) knowledge originates in the need to solve problems, (ii) efficiency in solving problems is the criterion for knowledge to survive. This fundamental link between knowledge and problems call for a co-definition of both, a way of research that situated learning invite to explore. To define and formalize what is a problem will not be easier than defining and formalizing what is knowledge. It has had less attention from research until now.

Let us consider the schema shown in Fig. 1.

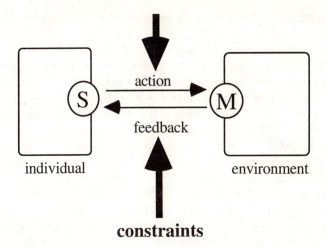

action

feedback

individual   environment

constraints

Figure 1

Instead of a direct description of or a definition of knowledge, it may be sufficient in our field to characterize it by its effects. Our main concern is to be able to ascribe some knowledge to an individual as the result of his or her activity during a session with an AET environment.

Knowledge can be characterized by the dynamical equilibrium of the loop action/feedback of interactions between a *knowing subject* S and a specific *milieu* M in the context of certain constraints. In this schema, the knowing subject is not the individual in all its complexity, but the individual from the point of view of the S/M interaction, and the milieu M is not the whole "exterior" environment of the individual but only a "relevant" part of it. M might be a physical device, or an individual (possibly the subject itself), or a social body, or whatever else. The constraints determine the specific economy of the system (Popper 1979, p. 78, Bourdieu 1980, p. 145).

I will not explore here the consequences of such a view of knowledge, I will use it to raise some of the questions still open in our field of research. The aim of AET is to implement a milieu M so that if the system (S, M) reaches some state of equilibrium it is acceptable to say that the individual knows some piece of knowledge. Given a piece of knowledge K:
– *what must be the characteristics of M?*

If K is a skill, the answer seems easy to find, we could for example rely on simulations opening the possibility of acting as if it were real situations, the learner getting negative feedback in the case of a wrong decision. Actually, it is not so easy since there is always a significant distance between a simulation and its reality of reference. If K is an abstract knowledge, in Ohlsson's sense, the answer is more difficult. It sends us back to Ohlsson's question: *what is the function of abstract knowledge?* Or, in our terms, what is the specificity of a milieu requiring the construction of what could be considered as an abstract knowledge.

– *how can a learner come to an interaction with a given milieu M?*

S and M defines the one the other. The existence of M for an individual is the result of a process which at the same time results in the construction of S as a knowing subject. There is here the origin of a paradox of teaching to which I will come back later, but it is already clear that it raises a problem whose complexity is not the same in the case of young learners and of adults.

– *what is the rôle of constraints on the interaction (S, M)?*

In particular, teaching and training situations determine their own constraints on the interaction. These exogenous constraints are likely to provoke the construction of unintended meanings. In the case of AET, the milieu is provided by the technology and interactions are under the constraints of its interface. Then, the representations chosen at the interface, as well as the internal representations which are in command of the whole functioning, are likely to play an essential rôle.

2.2 Computational Transposition: The Impossible Fidelity

Once some knowledge has been identified as an object of teaching or training, a series of adaptive transformations must be applied to make it teachable and learnable. This process which turns knowledge into an object of teaching is called the *didactical transposition* (Chevallard 1985). It is not a mere simplification which decreases the complexity of a given piece of knowledge while preserving its meaning. Because of the links it establishes with old objects of teaching, and because of the staging it needs in order to make teaching possible, didactical transposition is likely to modify the very meaning of the knowledge it works on.

AET introduces a new source of transformation of knowledge due to the fact that their design and implementation involve beliefs and conceptions of the so-called author – indeed usually a team – about learning and teaching. It is also due to the specificity and the constraints of sophisticated symbolic technology involved at both the interface and the "inside" of the AET environment, as well as technical constraints like those imposed by operating systems and hardware.

In the case of computer-based environments two types of constraints due to the implementation of knowledge representation are quite classical and very often mentioned. They are *granularity* and *compilation*. Knowledge compilation and granularity bind on the inspectability of systems and their capacity for the generation of explanations; obviously, the later is of a special importance from an educational perspective. These aspects of implementation are quite classical and often referred to by AI researchers. I would like here to point out other ones which could be of importance in our evaluation of computer-based learning environments and which receive less attention in the literature.

Important decisions of implementation are related to management of time and sequential organisation of actions, which implies the introduction of explicit order where, for most of the users, order is not normally expected or does not even matter. For example in geometry, Cabri-géomètre (Laborde J.-M. 1986, 1993), an environment allowing direct manipulation of geometrical object, induces an orientation on the objects: the segment AB is oriented because A is created before B. These artifacts prove to be coherent with the behavior of objects in direct manipulation, but contradictory with the fact that in paper-&-pencil these objects have no orientation – unless it is explicitly stated. An other decision might have been taken, but in this case the behavior in direct manipulation would have been at least counter-intuitive in a non productive way (Balacheff 1993). In general, specifications of such software bind in a fundamental way what they are able to provide as a milieu for the learning of geometry (Laborde C., 1992).

These questions are crucial when the knowledge concerned is related to time. For example, NEWTON (Teodoro 1992) provides learners with an environment in which they can observe and experiment with behaviors of particles in the context of elementary dynamics. Because of the management of computations the movement represented on the screen appears unexpectedly to speed up or to slow down; considering the time display allows the realization that the time is not uniform: the machine[2] takes time to compute. This software has some perceptual fidelity, but constraints of computing provoke some distortions which can be diminished but not completely suppressed – they are likely to appear again in limiting conditions.

The effect of such technical constraints are well known in computer graphics which deal with discrete screens. For example, the representation of the graph of a function requires a stroboscopic procedure (sampling of the value of the variable, choice of the points on the screen) which is likely to produce unintended phenomena in some cases (see for example the representation of the graph of sin(exp x)) for x>4 with software like Anugraph, Mathematica effects violate the WYSIWYE principle: "what you see is what you expect" (Bresenham 1988, p. 348), which in the case of learning environment might lead learners to develop unintended meanings.

The choice of a system of representation determines the kind of manipulation which can be made available to the user. Let's consider algebraic expressions like the ones we manipulate in elementary algebra, for example 5x+2x(x-3). These expressions can be viewed as strings of characters or as a tree structure. To chose the one or the other fixes the kind of manipulation possible at the interface of the system. If the list structure is chosen, which is the case for PIXIE (Sleeman 1982), then the "buggy" transformation 5+3x —> 8x is possible, whereas it is impossible with a tree structure (one cannot find any sub-tree manipulation which "explains it"), which is the case of APLUSIX (Nicaud 1992). Considering such differences, we must realize that APLUSIX and PIXIE do not have the same teaching objectives (PIXIE focuses on mal-rules diagnosis in elementary manipulation of

---

[2] Observed with Newton 0.5 on an Apple Powerbook 180.

algebraic formulas, whereas APLUSIX focuses on strategies in solving problems such as factorization).

Reacting to what might be seen as mere (but complex) technological limitations, people could suggest other implementations so that these defects no longer exist. But such suggestions miss the fact that other implementations would give rise to other "side" effects or unintended effects. The question is not to suppress them, since any *representation has productive effects*, but to be able to express in detail what they are and what could be their consequences in the case of learning environments.

Multiple representations are very often suggested as possible solutions in order to overcome such difficulties, but it supposes that one can exhaustively enumerate and describe representations related to a given piece of knowledge. A brief look, at the history of human knowledge, even restricted to science, or at the various representations standing out as landmarks along the development of learners, can convince anyone that the enterprise is condemned to failure. Another solution which has not been explored yet is to characterize the domain of validity of the chosen representation, and thus the domain of validity of the educational software itself. Such analysis needs a strong pluridisciplinary competence.

The problem of the relationships between a representation and what it represents is more often than not raised in the literature in terms of *fidelity*. This derives from the domain of software design which aims at the training of behaviors: "A highly realistic depiction of the device or environment and equally realistic techniques for interacting with this interface may be necessary for the student to understand and learn the domain. This belief has led to the development of flight simulators and maintenance training systems that are based on real-world devices, and that give students direct experience with both the devices and the domain problem in which they are used" (Miller 1988 p. 175). Research in this direction has led to the development of interfaces which reify abstract concepts like speed, energy or vectors in order to make possible their "direct manipulation".

Searching for direct access to concepts is no more than an attempt to pass over the fundamental problem of the relationships within the semiotical triangle signifier, signified and referent, made even more difficult in some cases because of the absence of any concrete referent.

Let us call *computational transposition* (Balacheff 1993) the process which leads to the specification and then the implementation of a knowledge model. Computational Transposition refers to the work needed to fit the requirements of symbolic representation and computation. Since it is not possible to find a solution avoiding bias between representations and that which they aim to represent, the way could be to delineate the *epistemological domain of validity* of the chosen modelisation or representation (Balacheff 1991, Balacheff and Sutherland 1994). An essential research issue for the coming decade is to understand the process of computational transposition, especially its intrinsic characteristics (the one which will not be modified by technical progress), and to develop theoretical frameworks and methodology for the identification of the epistemological domain of validity of AET.

2.3 Knowledge Specificity as a Constraint

The characteristics of the milieu for the learning of mathematics, physics or foreign languages are fundamentally different. The milieu for physics is part of the "real world", for foreign languages it includes human beings, for mathematics it is already a theoretical system. Although this observation seems obvious, most of the research projects claim that they contribute to the field at a general level and pretend to be domain independent. Even though, from a methodological point of view, they have developed no means and no theoretical framework to provide evidence for the validity of this claim. This tendency is very strong in the field of ITS research, as Verdejo points out: "Flexibility and adaptability to the student as well as independence of a particular teaching domain are important requirements in ITS design" (Verdejo 1992, p. 146). This position is probably due to the pressure of A.I research on the domain of AI and Education, following which ITSs appear as an application field of AI. To some extend, AI is not accountable beyond the coherency and computing robustness of the software it produces – one constraint often mentioned to justify the search for domain independence is re-usability.

There is a movement towards more "pragmatic" approaches. The idea is to develop limited but efficient environments, definitely bounded by a certain domain of knowledge. This direction of development, close to engineering, is likely to limit the progress of research although it might be politically efficient. Indeed it is essential to have efficient and reliable AET, but with the risk that this could be obtained by ad hoc design and implementation in order to fulfill the constraints of too narrow topics.

The problem of knowledge dependency is some what different and more fundamental. Its recognition starts from the fact that some domain of knowledge have their own specificity and characteristics. Let's come back to the case of mathematics and physics. The criterion for the validity of a statement in mathematics is some how internal to this discipline, I mean it relies on techniques of symbolic technology whose more achieved form is the "mathematical proof". In physics, the validity of a statement is dependent on the relationships between the theoretical models and an "experimental field". The "experimental field" is the "field of validation of the theoretical construction [… and …] it is the place where information is taken by direct perception of events or by measurements" (Tiberghien 1992, p. 195). This field is a theoretically organized piece of reality, but it leaves to perception – under the control of the theory – a central place. This difference between mathematics and physics is crucial. Such differences can be found between all the disciplines.

Domain independent theoretical frameworks and methods for the design and development of AET will not be able to take into account these specificity's and they may even try to avoid them with the risk of becoming serious obstacles for further developments.

What is domain dependent and what is domain independent in an AET system is an open question. Its solution requires an answer to the question of the nature of specific domain knowledge and of the related complexity of their learning and teaching, as well as an investigation of the way general cognitive structures adapt to the constraints of a discipline. Before ending this section, I would like to mention that the choice of a content to be taught as well as its specification are not a purely educational problem. It must take into account the fact that technology changes the nature of knowledge which can be made available to learners. It is now clear that graphic pocket calculators, as well as four operations calculators in the "recent" past, not only provides learners with new tools but also implies a transformation of the mathematics to be taught. The development of low-cost computers will in a near future make available powerful simulations, microworlds and tutoring systems to learners (children and adults as well). In particular, the development of "interactive electronic books" opens the way to a revolution close to the one of printed books in the dissemination of science and culture. Research on advanced educational technology must integrate the technological as well as this educational dimension in the investigation of the features of AET systems. This reflection on the content of teaching or training, under the light of technology must be investigated in each domain of knowledge taking into account its own specificity.

3   Modeling Issues

3.1 Modeling the Learner

Student modeling is the problem which has surely been in the forefront through out the 1980s. I rely on the scientific authority of one of the leading researchers on this theme, J. Self, who asserts that "the learner model is the central and yet most controversial component of intelligent systems" (Self 1990, p. 46). This might become even more controversial with the emergence of the situated epistemology.

The historical milestones in this domain are overlay modeling, the incorporation of specificity's of learner knowledge by the means of bug libraries or tentative models of bug generation, or bug reconstruction using machine learning techniques. Important difficulties have been encountered and are still resisting research effort. Different direction for progress are open, they are radically different, ranging from the search for psychologically valid models to no models at all. I agree with Self on the origin of this controversy, the roots are in what the tradition names the psychological plausibility of the models. However, the functions of student models in AET, and also their specifications, is not clear. Some suggestions can be made to try to clarify this question: if the AET essentially relies on dialogue and human like communication then a psychologically valid model – but within what limits? – is required, if it is essentially a reactive environment then it is necessary to take the user into account but the relevance of this model depends on the adequacy of the feedback it allows and not on its psychological validity *per se*.

The problem of taking account of the learner in the course of a session is two sided. On the one hand we must be able to characterize the kind of events which are relevant to make sense of the learner's behavior, on the other hand we need a function, usually named *diagnostic*, to ascribe some formalized meaning to the observed events. The following schema (Fig. 2) sketches the different elements involved in the problem of student modeling and their relations. Reading this schema, one must remember that the identification of "facts" is the result of a cutting out and of an organization of "reality" under the control of a theory and its related methodology. When researchers in education or in psychology create a corpus of observed "facts", its relevance depends on the theory and the methodology they use. Some of these facts are organized in order to describe what we call *behaviors*. In their turn, these behaviors are used as data to elaborate models – which I have called the *learners' conceptions*. It is clear that such models are the construction of researchers and not at all what is "in the students' heads". In the case when a computer-based system is the observer of the external "reality", the events which are likely to constitute the observed "facts" are physical events

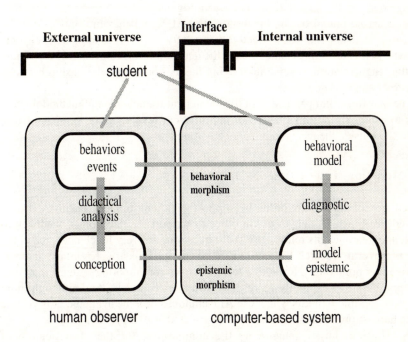

Figure 2

captured at the interface. Physics can provide us with a description, but even this level of description cannot be considered as "objective": it is a kind of organization of some "reality". Then, a *behavioral model* can be obtained from the analysis of the events occurring at the interface of the system, a criterion of validity would be the fact that when run in the same environment, it creates at the interface the same observable "facts" as those which observers have identified with reference to their own criteria. I call *behavioral morphism* the mapping between the behavioral model and the description of the student's behavior that researchers may produce; it is more than just the history of students actions at the interface: it must give an account of the structure of this action seen through the lenses of the system. An *epistemic model* can be constructed using as data the behavioral model, it is the result of what is usually called a *diagnostic*. I would consider that such a model is valid if there exists an *epistemological morphism* between it and the students' conception as elaborated by research, that is: a mapping which preserve the epistemological structures.

Considering the specific aim of AET, it is clear that there is a special need in this domain to take into account the user as a knowing system. Taking the categories of McCalla (1988), the answer could be either external student modeling or internal student modeling. But, external modeling does not mean the absence of a formal representation of some student knowledge in the system. Whether it is locally situated (as a specific module) or disseminated in various places in the code is not the question. The question is whether it is adapted to the need of the management of the interaction, the psychological constraints on an epistemic model must be bound by the function of the AET. A behavioral model may be sufficient, its data being used directly by the software in order to make decisions. But even at this level one cannot escape the question of the relevance of those data and thus their relation to some underlying theory about the relationship between the learner and the knowledge.

The question still open is that of the theoretical framework and methodological tools to establish the links between the different levels and types of modeling inside and outside the technology-based environments.

3.2 Modeling Interaction

Which model to use in order to support efficient tutorial or instructional strategies? Which model to use in order to support the decision of feedback to users, and guarantee its relevance? These questions are very much open and at the core of several research projects in the field. As Dijkstra, Krammer and van Merriënboer point out: "Nowadays, there is a need for authoring systems that offer the opportunity to explicitly represent instructional models. And although more and more flexible tool kits which use AI techniques become available, it is yet one of the hardest problem in the field." (Dijkstra et al. 1992, p. 2–3).

A criterion, largely shared by the community, for the classification of educational software is the degree of student's initiative which is allowed or equivalently their degree of directiveness. This leads to a classical classification from ITSs which leaves almost no freedom to the learner – Anderson's tutors are very much of this type, to microworlds which leave a complete freedom to the learner within the syntactical constraints of the software – LOGO is a famous archetype of such software. In between, one can find software referring to coaching or apprenticeship. Critics of the extreme cases, tutoring systems or microworlds, are well-known. For the latter there is evidence that with no external constraints, at least *given* tasks, it is unlikely that students would learn anything. Coaching systems are an attempt to find an intermediate solution. They leave students apparent freedom, which means that they do not react systematically nor immediately to student errors, but some functionality's allow these systems to interact following an evaluation of the students' conceptions in order to influence them. The paradigm of *guided discovery learning* (Elsom-Cook 1990) goes beyond this position, attempting a real balance of initiative, depending on an evaluation of the student's needs. It could provide very open environments, such as microworlds, for some purpose and then shift to tutor-like behavior if the situation of the learner is such that it seems better to do it, or if the teaching target appears to be better reached in this way.

Tutorial or instructional strategies are in control of the concrete decision to be made along the course of an instructional session. They have to answer very basic questions, like: When to react to the learner? Which reaction? Must learners be told about their errors? Or, must the situation be organized so that he or she can discover it? There are few cases in which the answer to these questions is based on a solid theoretical framework. Dijkstra and his colleagues mention some of them due to the already classical work from Collins, Anderson or Jonassen (Dijkstra et al. 1992, p. 2).

A direction of research aimed at solving this problem of modeling interaction consists of observing "good" teachers as experts. There are critics about this approach (cf. for example Self, etc.), which finally raise the question of the specificity of AET with respect to teaching. AET, as considered in this paper, aims at a one-to-one interaction oriented towards some learning. The teacher has a lot of other issues to consider, related to the management of the classroom society. Learning is one of his or her responsibilities, but the teacher is responsible for the socialization and the education of boys and girls in a larger sense. Teachers are part of a system which has quite a specific aim and economy, how this influences their professional behavior from the point of view of learning has not been clarified. In fact the question – to the best of my knowledge, has not yet been asked.

These considerations invite us to consider in different manner two different types of AET, one based on human-like interaction (like Socratic dialogue), and the other based on providing learners with an environment likely to facilitate their learning (the so-called constructive AET).

Let us come back to the evaluation of microworlds and tutoring systems:
– First, a free exploration of a microworld offers learners a rich experience but does not guarantee that a specific learning occurs.

- Second, the close interaction of a tutor could guarantee performance but not the nature of the underlying meaning. Learning in such an environment could mean learning how to obtain the best hints and help from the tutor so that the problem at hand can be solved.

Two ways can be explored to overcome these difficulties. The first one consists of extending the notion of microworld so that it includes the teacher and a larger environment of the learner. It aims at incorporating the means for regulating the learners' activities (Hoyles 1993). The second orientation consists of building environments specifically for the acquisition of a given knowledge K, a milieu for the learning of K. Such a milieu must have the following fundamental characteristics: (i) to allow the implementation of problems specific to K, (ii) to allow learners to consider these problem at different levels or with different initial solutions, (iii) to provide meaningful feedback so that the learner's solution can evolve towards the one giving rise to the new knowledge.

Microworlds are privileged tools to model such milieu. Starting with a few simple primitives, the learner can construct more and more sophisticated objects and define more and more complex tools for further investigations. In some sense, the microworld evolves as the learner's knowledge grows. Cabri-géomètre (Laborde J.-M. 1986) is a significant example of such a microworld. By means of macro-constructions, it enables the learner to build into the microworld new objects or new tools with a few elementary primitives as starting points. It shows clearly that modeling a milieu is closer to modeling knowledge, than to modeling the student. It needs deep epistemological analysis in both the historical perspective and the psychological perspective to define the proper ecology of the given piece of knowledge. For this purpose, external learner modelling is still necessary, but it does not mean that such a learner model must be embedded in the system at an epistemic level. It is necessary for the specification of the behavior of the computerized milieu, especially the nature of its feedback.

4 Situated AET

4.1 External Validation

The title of this section, "situated AET", could seem to be the price paid to fashion. In fact, it is both a wink and a relevant key word for what follows.

Let us take the case of the famous software WEST, the nature of the game it proposes is very different whether it is used at home during leisure time or in the context of a classroom activity. The context of the classroom means that the game played has some intended purpose, some agenda the teacher or the trainer has in mind. The different meaning of the same software could lead to different meanings constructed by students. To push a little, we can imagine that in one case the learner uses the software for his or her own pleasure, in the other case he or she does so for the pleasure of the teacher. We are close to the distinction introduced by Lave (1988), and borrowed from economics, between the use-value of knowledge and its value of exchange. The same idea of economy led Brousseau at the beginning of the 70s to link the meaning of knowledge to the characteristics of the situation in which it was developed, taking especially into account the proper economy of teaching/learning situations (Brousseau 1972, 1986).

In the case of AET, the use-value depends on its capacity to achieve efficiently its fundamental goal: to allow its user to learn a given piece of knowledge. This remark sounds like common sense, but it just reminds us of what should be the starting point for the question of evaluation in our field of research: knowledge.

Such an evaluation is not easy, it calls for tools to evaluate the learners' conceptions after an AET session. In his stimulating analysis of the difficulties encountered by ITS research, McCalla (1992, pp.115–117) points clearly to the fact that the answer to this question will not be the result of a mere transfer of methodology already available in education research, psychology or other AET related areas. The reasons come from two different sources: firstly the complexity of any real teaching or training situation makes hazardous the control of the evaluation in real situations, secondly the specificity of AET must be taken into account in the design of any evaluation. So, the solution will not come from other domain, their own methodologies could have some heuristic value, but this is not enough. The field must coin its own tools, means and methods. If we borrow some of them, at least we have to establish their relevance to our own objectives.

Several problems of the evaluation of AET have been examined by other researchers. I will here restrict my contribution to two points:
- the didactical contract, which refers to the meaning for the learner of an AET system in situation.
- the issue of socialization of knowledge, which is at the origin of some of the difficulties in implementing AET systems.

4.2 The Didactical Contract

Any situation which involves an AET system has an objective, known to the learner, to teach or to train. For this reason learners could always be tempted to guess what the system intends to teach them. For example, the "Socratic dialogue", the philosophy which underlies the design of several educational software, could provoke student behaviors closer to the guessing of the machine's expectations than to a construction specifically related to a piece of knowledge. Such behavior is classical and has been observed in the case of the use of menus: learners browse in order to try to guess what to do next instead of constructing the next step from the information at hand.

Here is one of the important paradoxes of teaching, especially with young learners: if you tell learners what you want to teach them you partly kill the source of meaning, if you don't you create what could appear to be a fake situation – since anybody knows that there is a hidden agenda. If we take the case of WEST, it is obvious that if you say which properties the software allows to be discovered you kill its real interest, but if you just play then the learners might try to guess what they must learn and what they think has been hidden in *the game as a didactical tool.*

The solution to this paradox lies in a negotiation between learners and teachers to preserve what we could call the "game of knowledge". In some sense, the teacher "asks" students to behave *as if* it were a genuine problem situation for them. Here the teacher has an essential rôle to play, negotiating a context likely to allow students to make sense of the interactions they will experience. The concept of *didactical contract* (Brousseau 1986) has been coined as a tool for the analysis of the relationships between teacher and students with respect to the meaning of the didactical situation. More often than not, the object of the didactical contract is not explicitly stated, it is a result of the social interaction in the classroom and its existence is revealed when it is broken by one of the partners involved. This contract may have an effect on the nature of the students' way of knowing since they can always try to find evidence of and to fulfill some teacher's expectation .

Then the introduction of educational software in the classroom raises complex problems, not only from the point of view of concrete management of space, time and machines, but also from the didactical point of view. Suggesting several scenarios, Vivet points out the complexity of the didactical management of a classroom in which teachers collaborate with computer based teaching-learning environments and delegate to them part of their own responsibility. He raises basic questions such as: How can a machine decide to pass the control to the teacher to solve a problem which is not of its competence? Or, in the case of collaborative learning, how could it establish some relationship between students having difficulties and other students who could help them? (Vivet 1992). From a didactical point of view, the context of knowledge functioning is quite modified, even complexified.

The teacher organizes the meeting between an educational software and a student. He or she decides when this meeting will occur and she states or not the reasons why it occurs in the larger context of the didactical process being engaged in. This position of the teacher is often seen as the Achille's heel of didactical situations. But, the fact that students interact directly with the computer without direct input from the teacher is not sufficient to guarantee the success. It is only by means of a negotiation which aims at the devolution to the student of the control of his or her interaction with AET provided to him, that the quasi-isolation of the student from his or her teacher could be obtained. The didactical contract ruling the situation will allow it to be close enough to a situation "free" of the teacher's influence. But indeed, students can always try to find in the behavior of the educational software indications of their teacher's expectations, and then try to complete the task without necessarily functioning within the intended epistemological register.

Educational Technology is only a media supporting the interaction between learners and a milieu specific to some learning purpose. This intentionnality might hide the epistemological function. This difficulty can probably only be overcome by means of a negotiation conducted by the teacher. For this purpose, how can a theory of AET in-situation be specified? Could such a theory be incorporated into the theoretical rational of the initial design?

4.3 The Necessary Socialization of Knowledge

Again, there is very often "in the air", the idea that AET will lead to the disappearance of schools and teachers. It may be for this reason that very little attention is paid to the possible interaction between teachers and AET, compared with the emphasis on learners' AET interaction. At least in the case of general education, this dream must be forgotten.

In the case of general education, teachers are necessary for basic reasons, which includes:
(i) to organize the learning of young people who are far from being able to identify what is worthwhile to learn from our "informational era".
(ii) to guarantee the socialization of knowledge and the "compatibility" of that knowledge with that of the world outside the classroom.
(iii) to allow students to recognize among all their productions, those which are to be kept as "knowledge"; that is to "institutionalize" knowledge.

This last reason is very often forgotten, although it is essential for "teaching" to take place. Among all the intellectual constructs to appear during the class activity, both at a collective and at an individual level, students cannot imagine which ones will be important for future use within or outside the classroom – i.e. the ones which are not only locally relevant, ad hoc to the current situation, but part of a new knowledge (Brousseau 1986, p. 71). Actually, this problem is not just the problem of general education, it is also the problem of any training situation in which much is left to the activity and initiative of trainees, as in the case of apprenticeship. This need for the socialization of knowledge, implies the possibility of a collaboration between AET and teachers or trainers which goes beyond the mere transfer of some of the teaching or training task to some technology.

Teachers will not be fully able to put AET into practice in their daily life, if they are not well informed about all the aspects which could determine its place and its precise role in the teaching process. To some extent, we can say that they must know it from a professional point of view, as they would like to know a colleague with whom they might have to share the responsibility of their class. Indeed this raises the question of the specification and of the information about the knowledge to be "encapsulated" in the machine and the way it functions with respect to a given learning/teaching target. But it also raises the problem of the communication between the teacher and the machine about the learning process. In the case of artificial intelligence, much research has dealt with the problem of explanation generation for teaching purposes, but all of them are centered on the student, none on teachers and the explanation they might expect about a didactical interaction.

The machine must be able to handle and produce relevant didactical information about the teaching process, in order to be able to interact and cooperate properly with the teacher.

5   Conclusion

There is in the tradition of AI a classical criterion for the validation of a model: the Turing Test. In the case of educational software, the Turing Test might become: let a computer together with a teacher, be both hidden from the view of some observer. The observer, by just putting probing questions to each of them, has to try to decide which of the two is the computer and which is the human being. If the observer is unable to identify the human subject in any consistent way, then ... But if the observer believes that the machine behaves like a teacher, will we consider that it means that this machine is adequate for the intended learning objective? Surely not. The problem is not to recognize that a "teaching" machine behaves like a teacher, but rather that it provides feedback allowing one to construct a "correct" meaning for some knowledge it intends to teach.

The fact that the Turing Test becomes meaningless in the case of education invites us to leave the anthropomorphic view of the scientific program of Advanced Educational Technology, to the benefit of an epistemological view.

Our aim is to create the conditions of an artificial knowledge growth ecology, creating environments allowing efficient learning under institutional constraints. The fact that some environments would mimic teachers is just a matter of economy: it was easier to look at the only experts we have at hands. But looking closely at the object of this expertise invites the realization that its object, acquired through practice, is only partly related to knowledge under the constraints of teaching or training systems. For a large part it is related to education in a wider sense, which is in my opinion not within the scope of AET.

Thus, it is from the point of view of the epistemology of teaching and training systems, that we have to consider our field of research and to state its specific questions. Knowledge and learning not being considered "in general" but under the specific constraints of these systems.

Acknowledgments

Several versions of this paper have benefited from rich discussion during panel meeting. I would especially like to thank Gilbert de Landsheere, Philippe Duchastel, Michael J. Spector and Rosamund Sutherland, who wrote me stimulating and helpful comments.

References

(NATO ASI Series F volumes are indicated by F and volume number)

Balacheff, N. (1991) Contribution de la didactique et de l'épistémologie aux recherches en EIAO. In: Bellissant C. (ed.) Actes des XIII° Journées francophonessur l'informatique. Grenoble-Genève. 9-38. Grenoble: IMAG

Balacheff, N. (1993) La transposition informatique. Note sur un nouveau problème pour la didactique. In: Artigue, M., Gras, R., Laborde, C., Tavignot, P. (eds.) 20 ans de didactique des mathématiques en France. Grenoble: La Pensée Sauvage

Balacheff, N., Sutherland, R. (1994) Epistemological domain of validity of microworlds, the case of Logo and Cabri-géomètre. In: Lewis, R., Mendelshon, P. (eds.) Lessons from learning. Proceedings of the IFIP WG3 working group A46, 137-150. Amsterdam: North-Holland/Elsevier

Bourdieu, P. (1980) Le sens pratique. Paris: Editions de Minuit

Bresenham, J. E. (1988). Anomalies in incremental line rastering. In F40

Brousseau, G. (1972). Processus de mathématisation. In: La Mathématique à l'école élémentaire. 428-442. Paris: Association des Professeurs de Mathématiques de l'Enseignement Public

Brousseau, G. (1986). Fondements et méthodes de la didactique des mathématiques. Recherches en didactique des mathématiques 7(2) 33-116

Chevallard, Y. (1985). La transposition didactique. (Nouvelle édition revue et augmentée, 1991). Grenoble: Editions La Pensée Sauvage

Clancey, W.J. (1993) Guidon-Manage revisited: A socio-technical system approach. Journal of Artificial Intelligence in Education 4(1) 5-34

Dijkstra, S., Krammer, H.P.M., van Merriënboer, J.J.G. (eds.) (1992) Instructional models in computer-based learning environments. F104

Elsom-Cook, M. (1990) Guided discovery tutoring. Chapman, London

Hoyles, C. (1993) Microworlds/Schoolworlds: The transformation of an innovation. In F121

Laborde, C. (1992) Solving problems in computer based geometry environments: the influence of the features of the software. Zentrablatt für Didactik des Mathematik, 92(4) 128-135

Laborde, J.-M. (1986) Proposition d'un Cabri-géomètre, incluant la notion de figures manipulables. Sujet d'année spéciale ENSIMAG

Laborde, J.-M. (1993) Intelligent microworlds and learning environments. In F117

Lave, J. (1988) Cognition in practice. Cambridge, UK: Cambridge University Press

Lesgold, A., Katz, S., Greenberg, L., Hughes, E., Egan, G. (1992) Extensions of intelligent tutoring paradigms to support collaborative learning. In F104

McCalla, G. (1988) Intelligent tutoring systems: Navigating the rocky road to success. In F96

McCalla, G. (1992) The central importance of student modelling to intelligent tutoring. In F91

Miller, J.R. (1988) The role of human-computer interaction in intelligent tutoring systems. In: Polson, M.C., Richardson, J.J. (eds.) Foundations of intelligent tutoring systems, 143-189. Hillsdale, NJ: Lawrence Erlbaum Associates

Nicaud, J.-F. (1992). A general model of algebraic problem solving for the design of interactive learning environments. In F89

Ohlsson, S. (1988) Towards intelligent tutoring systems that teach knowledge rather than skills: Five research questions. In F96

Popper, K. (1979) Objective knowledge. Oxford, UK: Oxford University Press

Self, J. (1990) Theoretical foundations for intelligent tutoring systems. Journal of Artificial Intelligence in Education 1(4) 3-14

Self, J. (1992) Computational mathetics: The missing link in intelligent tutoring systems research? In F91

Sleeman, D.H. (1982). Inferring (mal) rules from pupils' protocols. Proceedings of the European Conference on Artificial Intelligence 160-164. Paris: université d'Orsay

Smith, R. B. (1988) A prototype futuristic technology for distance education. In F96

Teodoro, E.D. (1992) Direct manipulation on physical models in computarized exploratory laboratory. In F84

Tiberghien, A. (1992) Analysis of interfaces from the points of view of epistemology and didactics. In F86

Verdejo, F.M. (1992) A framework for instructional planning and discourse modelling in intelligent tutorial systems. In F91

Vergnaud, G. (1992) Conceptual fields, problem-solving and intelligent computer tools. In F84

Vivet, M. (1992) Uses of ITS, which role for the teacher? In F91

2

Advancing Education by Proper Technology

D.G. Bouwhuis, R. van Hoe, H. Bouma

Institute for Perception Research/IPO
Eindhoven, The Netherlands

Abstract: The major trends emerging from a review of the first 25 volumes in the NATO Series on Educational Technology are a convergence on Intelligent Tutoring Systems and on microworld simulation tools. In both approaches a radical reform of the entire educational system is generally favored. In this paper the relative merits and drawbacks of these systems are critically reviewed, also with respect to the educational context in which they will have to be operative. The general conclusion is that both empirical validity and evaluation are seriously lacking. This picture does not hold for the field as a whole, however. Though small in number, various authors produce fascinating and new accounts of learning processes made observable in interactive learning environments, of developing mental models, of learning strategies and social interaction in learning. Recommendations for future research go in this direction and stress the importance of fundamental research on learning, systematic student-centered evaluation and the design of effective instructional dialogues and interfaces. The fact that the most successful computer-based instructional system to date, the flight simulator, has not been mentioned once in the series provides an interesting case in this respect.

Keywords: Intelligent tutoring systems, microworld simulation tools, interactive learning environments, student centered-evaluation.

Introduction

Learning researchers share two certainties. The first one is that learning by itself is intrinsically good; the second is that the prevailing educational system is structurally wrong. These convictions are not recent; they have been held for centuries and seem to gain weight gradually over time (Tinker & Thornton, 1992; Frederiksen & White, 1992). In the NATO AET series this educational concern, shared by various disciplines, has apparently led to a surprisingly general convergence of ideas about educational means. But though convergence has the appeal of desirability, in the current series it cannot dispel the notion of narrowness. A closer look also conveys the impression that cooperation between disciplines, usually highly acclaimed, may just as well produce the wrong as the right results, since the vices of the approaches involved may be combined rather than the virtues. The problem is that what may have been a quality, or even a virtue in one specific discipline, may turn into a vice when it is combined with concepts from a different discipline. Consequently, rigidly held beliefs and convictions may preclude a critical reflection on what is outside the direct field of specialization. Not all is bleak, however. Demonstration and simulation possibilities have soared, permitting realistic rendering of phenomena, while interactivity has been boosted to previously unattainable levels. Recently, the actual use of educational systems has spawned inte rest in the nature of mental models that students develop about (mostly) physical processes. The great variety of such models renders a much clearer picture of the nature of understanding and misunderstanding than was available in traditional delivery teaching. Also, interest has evolved on learning strategies in unconstrained settings, drawing attention to the social context. Similarly, the importance of the learning context, or situation, on unsupervised learning is beginning to be realized. Yet, the latter are still rel atively few and isolated branches in the field with high hopes and few solutions. In this review we will try to broaden the framework of discussion by first treating the position of ET in society in general. Next follows a critical discussion of current educational systems in which we will focus on the critical assumptions and the intended goals. Finally, we will review the successes and failures of ET against the background of educational practice and an interdisciplinary view of human learning.

1   Educational Technology and Technology Trends in Society

If technology can be described as providing technical options for human purposes, educational technology is technology provided for educational purposes, i.e., for learning. The aim of educational technology is, therefore, the general availability in society of technical means that help individuals and groups to learn desired skills of physical, perceptual and cognitive nature. Much of present technology stems from micro-electronics. Also educational technology is often considered as some type of computer with hardware and software, visual and auditory displays, keyboards and mouses, sometimes termed multimedia. The term "advanced educational technology" is then perhaps too easily narrowed to present-day digital electronics as appears in personal computers or workstations with some user interface and a lot of software. It is true that such technology has rapidly pervaded society, particularly industry, research & development laboratories, and offices, but less so schools and homes. But has technology greatly helped us to reach desired educational goals? In the education for scientific research, the answer is affirmative, and this will also be the case for education directed at jobs in which the use of computer and software is a requisite. Examples can be found in word processing products in offices, in certain military training programmes and in the use of flight simulators for the training of pilots. The general education such as in primary, and secondary or even tertiary schools has so far been hardly affected, if at all, by such advanced technology. In heavily subsidized innovation programmes PC's have indeed been introduced at schools, but any massive use for education has not yet materialized. It is probably fair to state that copiers, video recorders, overhead-projectors and perhaps the telephone have had greater impact on education in society so far than have PC's, workstations and educational software. So why have hardware and software technology so far failed to fulfill the high expectations as to their serving massive education inside and outside schools, or put it the other way round: what interactive technology can really serve learning purposes?

1.1 The Role of Educational Research

It is here that scientific research may come in because it is aimed at providing insight by directed efforts. The type of research involved should primarily be educational research, oriented towards making the technology serve educational goals. The research should make us understand what learning process can be served by interactive media, what types of software and software architecture will support the learning process, how a proper communication between (student) user and (tutoring) system can be guaranteed; how the students can be held absorbed in their tasks as for example has been achieved in certain video games, and what the role of the teachers should be. As a final proof, the research should answer what learning can be achieved in sessions in which the students actually use the system individually or in groups and with or without supervision. It may be illuminating to dwell for a moment at the role that research has to play in the penetration of technology into society. First, progress in scientific and technological research will provide new technology options. What the proper role of such technology will become in society, is usually difficult to predict. In particular the early expectations at the time of introduction of new technologies are often quite different from any later massive applications. This amounts to stating that the future is notoriously difficult to predict and that only in retrospect it will generally be possible to select the few people that have been proven right out of the many that have been proven wrong. Once it is there, technology is not influenced automatically by application oriented research. On the other hand, educational purposes have a very high priority in society and so we would be unwise if we left the penetration of educational technology into society to trial and error. Properly oriented research in the disciplines of educational science is necessary to help steer developments and to keep technology in its role of serving proper educational purposes. Of course, such purposes need to be carefully defined, the notion of usefulness operationalized. Apart from application-oriented educational research there will also be room for pure educational research, i.e. disciplinary research for the purpose of advancing insights without any intention or direction at applications. The quality of such research should be derived from new insights and not from potential applications. Such applications might result, nevertheless, but the time scale involved will generally be unknown. Confusion may arise if such pure research, not intended for any medium term fruits, is directed at subject matter stemming from the world of applications. A careful description of the problems and goals to be addressed will help to avoid unjustified expectations.

1.2 Education and Socio-Political Infrastructure

The penetration of educational technology into society is not achieved automatically, even if systems of proven usability could be defined. Large scale applications require extended infrastructure, that can absorb and support the intended use (Grandbastien, 1993). The general availability of proper educational software on the various subject matters to be learned is certainly necessary, but substantial additional infrastructure is needed in widespread teacher education such that teachers learn how to give educational technology its proper place in the curriculum, and to find teaching schemes in which there is sufficient room for any proven contribution s from technology. The infrastructure needed is not only of technical or organizational nature. Political and social commitments need to develop for a controlled introduction of educational technology in primary, secondary, tertiary, and permanent education, including a great many special courses such as for immigrants (language courses, labor skills). There is a risk that technology push with insufficient educational base will penetrate society in the absence of a proper research effort directed at timely evaluation and pilot projects for useful early feedback.

2 The Special AET Programme

The above framework is a general one which should be substantiated by elaborating the state-of-the-art in the various instructional or educational sub fields. However, there is also room for some general concern since the unsatisfactory mismatch between expectations and reality turns out to have existed already for a great many years. The analysis of Jones (1992) going back to the beginning of the special NATO program on AET reflects a similar atmosphere of dissatisfaction: "we have not paid a sufficient amount of attention to the instructional capabilities of instructional software". Also the replacement of good old Computer-Assisted Instruction (CAI) by apparently more potent "Intelligent Tutoring Systems (ITS)", has failed to solve the problem of why such systems did not fulfill the high expectations of the field. Next, the problems were seen to focus on insufficient instructional design, in particular on insufficient representation of the curriculum as the goal structure for instruction and on studies concerning implementation and evaluation (Jones, 1992). It is therefore apt to try and analyze what changes have occurred in the time frame of the Special Programme on AET and reflected by the proceedings of the Special Programme.

2.1 Educational Technology and the Topics of Teaching

It can be noted that the study of instruction and learning is frequently motivated by political, cultural and social, and science training considerations, generally aiming at rather global teaching goals, but with a cavalier disregard for the means by which learning takes place in the human student. The topics and skills to be learned, chosen for study in the educational technology community often reflect this 'social desirability' of learning. Though human beings start right after their birth to learn numerous complex things, faces, voices, surroundings, directed movements, and, in due time, much more complex skills like walking, eating with a fork and spoon and tying shoelaces, we fail to spot these protracted learning processes in theoretical treatises. It is also undeniable that many people learn bad things they ought not to learn at all; not all learning leads to socially desirable results. It is obvious that learning has a powerful autonomous component, but the relative lack of explicit training makes it less interesting for educational technology. Yet, it would seem unwise for educational technology to neglect the impact of autonomous learning capability of people, inasmuch it proves to be effective and durable, and requires only minimal supervision. Educational technology and its associated research seems to concentrate almost exclusively on the learning of cognitive skills and the acquisition of abstract and declarative knowledge, shortly, the more interesting higher mental activities. As a rule they include language training, and the regular fields at high school, except physical education and arts. Two other important fields are topics from computer science and diagnostics and maintenance of complex systems, the latter mostly in a military context. These are certainly not the only activities that enable humans to survive, to pursue a successful career, or lead a satisfying life. However, what these learning activities have in common is that they, ordinarily, require teaching or supervision and guidance, without which the corresponding skills would never successfully develop. So, it becomes clear that whereas the word "learning" has probably the highest frequencies of occurrence in the series, the essence of the workshops has really been "teaching". Throughout the discussions that follow we will try to make clear that this is not a subtle distinction.

2.2 Advanced Educational Systems

The human activity boasting the longest standing controversy regarding its education is learning to read (Chall, 1967, 1979; Mathews, 1966); "reading machines" date from the beginning of the 19th century. Concern about general reading performance is still as high as ever. But even nowadays tutoring systems for reading as a rule are not "intelligent", they are very restricted in scope and surprisingly rare in the literature. This is true despite the fact that the required domain knowledge is not very complex or extensive. Despite this early head start, current introductions on educational systems start invariably in the late fifties of this century with the early trials using language laboratories, move on to programmed instruction, and, with computer-assisted (-based, -managed) instruction as way stations, attain the intelligent tutoring system (ITS), thereby implying the enormous advances being made in the field. Down the bottom of the line are still the drill-and-practice programs, direct descendants of Computer-Assisted Instruction (CAI), and nowadays mostly called Computer-Based Training (CBT). A relative newcomer, usually located at the top, is the "microworld" approach, sometimes derived from ITS architecture's, and which gives opportunities both for exploratory learning, and for simulation. The term Computer-Enriched Instruction (CEI) is also employed for this kind of instructional tools. Finally, a very different way of training is "distance learning" which, in fact, shares nothing with the other systems, but which can only be realized with a computer, or rather, a lot of them.

3.   Types of Educational Systems

3.1 The Intelligent Tutoring System Approach

The main problem with traditional computer aided instruction (CAI) systems was that they were unable to provide adaptation or individualization. In response to this and other problems CAI faced, Carbonell (1970) argued that the CAI problem could not be solved with out the use of artificial intelligence (AI) techniques. In the same line, Self (1974) argued that an interactive learning program should contain knowledge of how to teach, knowledge of what is being taught, and knowledge of who is being taught. Consequently, researchers applied an AI perspective to the problem of creating learning environments and CAI systems evolved into what now is usually called "Intelligent Tutoring Systems" (ITS). Although different architecture for ITSs have been proposed, there is considerable consensus in the literature that ITSs consist of at least four basic modules: 1. expert knowledge module (i.e., what to teach); 2. student model module (i.e., whom to teach); 3. tutoring module (i.e., how to teach); 4. communication (user interface) module. Though the NATO Special Programme on AET has provided a multifaceted overview of the merits of the ITS approach, it demonstrates at the same time that there are also serious problems connected with it.

3.1.1 Student Models

As current interactive learning environments are developed according to an expert system methodology, the expert module is the most central component of those systems (Frasson, 1992). A first consequence of this knowledge-based approach is that the student is mode led in terms of the representation model of the expert module. This has several implications: A central assumption of ITSs is that the cognitive diagnosis of a learner's errors is an essential step towards meaningful individualized tutoring (Mandl & Hron, 1992). The student model is mandatory to the goal of cognitive diagnosis, and consequently to the goal of adaptation. More specific it is assumed that: 1. student modeling allows feedback adequate to correct any misconceptions or missteps that result in error; and 2. beyond error correction, a dynamic representation of the developing knowledge of each student determines what to teach next and how to teach it (Snow & Swanson, 1992).

There are two major student modeling techniques: the overlay and the process model. In the overlay model the student is represented as a subset of the expert: the expert model plus a list of knowledge units the expert has but the subject does not. The process mode l approach has been proposed for the first time by Self (1974). Self argued that a student model should be executable, so that the student model simulates the cognitive processes by which the user solves the task. In the BUGGY project (Brown & Burton, 1978) it was found necessary to enhance the process model with incorrect or buggy rules so that the BUGGY system was able to replicate the error performance of students. Consequently, current process models use a library of predefined bugs to test what combination of correct and incorrect rules produced the observed error. The overlay model was very common in the early days of ITSs, whereas currently the process modeling approach on the basis of an error library dominates the ITS research (e.g., Hron, Bollwahn, Mandl, Oestermeier, Tergan, 1992). Summarized, both student modeling approaches, overlay and process model, model the student in terms of the domain expert, either in terms of what knowledge units the expert has but the student does not (overlay model) or in terms of what knowledge units (i.e., buggy rules) the student has and the expert not (process model). In line with the expert system approach, it is assumed that both types of student models can be executed by an inference engine.

3.1.2 The Empirical Status of the Student Model

Student modeling is not based on explicit knowledge of the (interactive) learning process. Very few learning systems take the history of the student's performance into account in modeling the student. Although some ITS researchers (e.g., Anderson, Boyle, Corbett, Lewis, 1990) claim that the design and implementation of their ITS is based on a cognitive model of learning, this is only partially true. The LISP-tutor of Anderson et al. (1990), for instance, uses a process student model in which learning is represented by modifications of the student model. This implies that the learning process of the student is only monitored in an indirect way. This issue was also raised by Shuell (1992) as he argued that most instructional systems embody an implicit model of learning. The difficulty with these implicit models is that they are a) intuitive and implemented with little awareness of their validity and b) based more on personal opinion or philosophical belief than on a thoughtful consideration of the large body of psychological research on learning. Thus, an understanding of how students learn, the relation between instruction and learning, and various ways in which these concerns can be incorporated into an ITS is essential in the design process (Merrill, Li & Jones, 1992).

There are many reasons why student modeling should be conducted in terms of the cognitive model of the student, and not in terms of the expert module. In the ITS literature (e.g., Wenger, 1987) it is often argued that the expert module should be transparent, so that each reasoning step can be inspected and interpreted by the student. This glass-box approach is frequently applied in computer-aided language learning systems (e.g., Rypa, 1992). However, it is not at all obvious that the representation model of the expert module corresponds indeed to the cognitive representations of the student. There is empirical evidence that expert knowledge structures do not provide the most useful models for teaching (McArthur et al., 1988; Roschelle, 1990). The development of the ITSs GUIDON and GUIDON2 for medical diagnosis is interesting in that respect (Clancey, 1981, 1987). The ITS GUIDON is based on the expert system MYCIN(ES), which proved to be inefficient because the explanations of MYCIN's rules were hard to understand for the students (Frasson, 1992). The reason was that the explanations were given by the expert module but without consideration for the medical students who reason in a different way. This led to a reconfiguration of GUIDON into GUIDON2 which had a more appropriate representation model for tutoring purposes. This discussion illustrates that although the ITS approach correctly emphasizes the importance of knowledge representation in computer-based systems, this issue is dealt with from an AI point of view rather than a cognitive point of view.

3.1.3 Forms of Cognitive Representation

In most ITSs a semantic/propositional and/or rule-based/procedural representation formalism is used. An immediate implication is that these representation formalisms constrain the type of domains, the type of tasks, and the type of learning an ITS can model, in view of the empirical evidence for other formats of knowledge representation. Experimental evidence indicates the existence of the analogical/pictorial type of representation in our cognitive knowledge base (d'Ydewalle, 1992). Modeling analogical knowledge is important in view of the increased research interest for "mental models" (Gentner & Stevens, 1983). Mental models are based on analogical representations, and function in a large number of problem situations. The importance of mental models is recognized in quite a number of intelligent learning environments for the domain of physics (e.g., Frederiksen & White, 1992; Hron, 1992; Reimann, 1992; Soloway et al., 1992;

Tiberghien & Mandl, 1992; Vosniadou, 1992). However, as these ITSs simulate mental models using rule-based systems, the question arises whether this symbolic representation formalism is sufficient to grasp the specific nature of analogical knowledge (d'Ydewalle, 1992). Inherent to the rule-based representation format is also the assumption of an addressable memory storage system. Recently, arguments against the assumption that intelligent systems require an explicit representation of knowledge have been raised in the connectionist and situated-agents approach. Research within connectionism or parallel distributed processing approaches indicate that a representation could be distributed over many storage elements rather than one (MacWhinney, 1992). Research within the situated-agents approach demonstrates that autonomous systems can be constructed which exhibit robust and variable behavior without the need of an explicit world representation (Bosser, 1992; Connah, 1992; Norman, 1993; Stucky, 1992).

There is indeed increasing evidence against the idea that production rules form the units of knowledge, and consequently against the idea that the production system approach can yield a psychologically valid model of information processing. Experimental research of Gluck and Bower (1988a, 1988b), Nosofsky, Clark and Shin (1989) and Nosofsky, Kruschke & McKinley (1992) showed that the acquisition and representation of well-defined concepts could be better predicted in terms of a (combined) connectionist and examplar model rather than a traditional rule-based model. Similar results are reported by McSpadden (1989) within the context of a complex problem solving task (i.e., diagnosis) (cited in Hunt, 1989). McSpadden found that the problem solving and learning behavior of her subjects could not be described in terms of the usage and induction of rules, but in terms of an exemplar-based model of problems and learning. According to this model, learning consists in storing the individual problem situations in long-term memory (LTM), and problem solving is described as the abstraction of similar problem situation in long-term memory. An observation related to this discussion is the fact that instructional measures derived from elaborate analyses of student protocols sometimes show zero or even negative effects during training (e.g. Hron, 1992). The fully rational measures apparently conflict with the actual perceptual/cognitive behavior of the student. A study of Lundell (1988) illustrates this observation in an elegant way (also cited in Hunt, 1989). In his PhD dissertation Lundell (1988) studied an artificial diagnostic situation. Lundell first developed a rule-based system for representing student knowledge. This was done by analyzing the results of a structured interview about the diagnosis task. This information was used to develop rule-based expert systems tailored to the (asserted) knowledge of each student. After this knowledge-elicitation phase, the students were tested on new diagnosis cases. The students were correct 75% of the time. The new test cases were also diagnosed by the expert system programs Lundell had developed. The programs produced the correct diagnosis for 55% of the test cases; well above chance but significantly below human performance. Perhaps even more telling, the rule-based student model extracted from a given student was no more predictive of that student's performance than were systems extracted from the responses of other students (Hunt , 1989)! Lundell also constructed connectionist models for each student. The models were feed-forward models with a single layer of hidden units, trained with the back-propagation algorithm. These connectionist systems were used to classify the test cases. Seventy-two percent of the test patterns were correctly classified, averaged over simulations. This compares to 75% correct classifications, averaged over students. Furthermore, on a system-by-system basis, there was a higher correlation between the responses made by a system and the responses made by the student from whose data the system had been generated, than between system responses and the responses of other students. Quite clearly, the connectionist models more accurately portrayed student behavior than did the rule-based models. This was true in spite of the fact that the rule-based systems contained the rules that the student had generated themselves. Extensive knowledge engineering did not produce a satisfactory model of diagnostic performance, but connectionist training did.

A final consequence of the procedural representation is that ITS researchers try to develop an adaptive system based on a top-down approach rather than a bottom-up approach. O'Shea (1979) has argued that one of the most important aspects of any CAI program is its response-sensitivity: an interactive learning program is more response-sensitive than another when it is more adaptive to the individual learning needs of the student than the other system. In a classical ITS system the students performance is interpreted within the representation model of the expert model. Hence the degree of response-sensitivity of the ITS is determined by the extent in which the set of all possible behaviors of the student fits within the representation framework of the ITS. As the student's performance is rarely consistent and it is in fact impossible to predict the full range of student behaviors, the response-sensitivity or adaptability of ITSs is limited. In other words, classical ITSs are not quite robust as they cannot cope with unexpected, i.e., not predefined, behavior of the student. Moreover, as further instruction is based on the hypothesized knowledge state of the student and not on the student's behavior itself, problems can arise because the student's performance was misinterpreted within the expert module .

As already mentioned, in an ITS the cognitive diagnosis of a learner's errors is essential for the aim of adaptive training. This approach requires that a great deal of information about errors and misconceptions for a given domain and a given population of students must be gathered (e.g., Hron et al., 1992; Ploetzner & Spada, 1992). One of the problems in cognitive diagnosis is to identify whether an error is the result of applying incorrect knowledge – the subject has made a "mistake", or whether the student intended to perform the appropriate action, but failed to do so – the subject "slipped" (Norman, 1981). A recent empirical study (Payne & Squibb, 1990) of elementary algebra errors nicely illustrates the problems associated with the cognitive diagnosis methodology: 1. the frequency of errors was skewed: there were many infrequent errors and few frequent ones; 2. errors were unstable: students made errors in an inconsistent way; 3. there was also a high degree of inconsistency between students: different errors had different explanatory power in different student groups; 4. it was almost impossible to make a clear distinction between slips and mistakes. Summarizing, cognitive diagnosis also suffers from the same problems mentioned in the previous paragraphs: cognitive diagnosis is not robust as the ITS tries to interpret the student's errors in terms of a predefined library of bugs or mal-rules. As empirical research shows that it is very difficult to classify errors in a systematic way, it can be predicted that "unexpected" errors occur frequently with which the ITS cannot cope.

3.1.4 The User Interface

The interface of an interactive system should be capable to deal with the entire range of the human interactive repertoire, which consists of many layers of abstraction and which is represented in different modalities. Any abstract cognitive concept must ultimately have its perceptual or motor representation in some form, that can be mediated by the interface. Also, the tutoring system must engage in an educational dialogue that conforms to the natural conventions in human dialogues (Beun, Reiner & Baker, 1994; Taylor, Neel & Bouwhuis, 1989). Though in some papers there is some ho-hum reference to the importance of an effective and understandable instructional dialogue, there is nowhere any evidence that dialogue theory is familiar for designers of ITSs. Not surprisingly, many ITS programs can be made to crash embarrassingly easily by casual users, showing that robustness down to the command level is not a serious issue in implementation. Frequently, prompts, requests and screen operations are so context dependent and generic that well-disposed students are unable to understand what they are supposed to do at a choice point. Designers tend to consider these problems as trivia to be settled in a final implementation (Ennals, 1992). All the same, it is far more the effectivity of the dialogue that mediates the impression of intelligence than what is firmly hidden behind it. Likewise, the graphical screen information of programs reflects the background of the designers; few screen lay-outs are shown in the ARW papers, but where they are shown, they feature a complex multitude of windows providing status information which is probably relevant for the designer's, but must be bewildering to an aspiring student. One of the results of the expert system methodology is that in the design process of an ITS the implementation of the expert system module is emphasized, mostly at the expense of the interface. As a result of this the evaluation research is neglected or often not considered at all. In our view the design of interactive learning environments (ILE) should be based on results of evaluation research. In analogy with Norman and Draper (1986) we could call this methodology the user-centered design of interactive learning environments. This approach implies that at all stages the design process of an interactive learning system should be based on 1. fundamental research on the learning process the system will support (see also, McDermott, 1992), and 2. continuous user testing with the prototype (see also Zucchermaglio, 1993). The first type of research is mainly of importance for student modeling purposes, whereas the second type of research evaluates the interaction between the instruction environment and the learning process. An often neglected fact in ITS research is that an interactive learning environment creates a task context (instructional dialogue, presentation, user interface, etc.) for the user which determines how optimal learning can take place. The user-centered design approach has been applied in the Reading Board project at the Institute for Perception Research/IPO (Spaai & Ellermann, 1990), where there was hardly a theoretical idea about initial reading at the start and consequently development took place largely on the basis of results of evaluation studies.

3.1.5 Software Engineering of ITSs

Continuing the discussion of the last paragraph, it is surprising that – except for Evans & Patel (1992) and Kaye (1992) – in the NATO ARW workshops little attention has been devoted to the software engineering (SE) aspects of developing ITSs and interactive learning systems in general. Although the development of ITSs can be based to a certain extent on existing knowledge engineering tools such as rapid prototyping and the KADS methodology (Wielinga, Schreiber & Breuker; Winkels, 1992), it is clear from our discussion that a special SE methodology for developing interactive learning systems is needed. Current authoring tools are not sufficient as they tend to emphasize the implementation phase and neglect the analysis and design phase of the software life cycle (Witschital, 1992). Ideally, what is needed is a CASE-type of tool which incorporates the principles of user-centered design and supports the whole life cycle of a computer-based learning system. Turning to reality, experience learns that ITSs are very complex systems with a (too) long life cycle. Similar to other software projects, courseware and training programs developed (at least partly) by the end-customers appear to be just more effective, durable and simple than programs developed by external vendors (Gayeski, 1992). Due to their complexity, ITSs require a lot of computing resources. Although one could perhaps argue that future hardware architecture will solve that problem, current ITSs can have long response times. It is a well-known fact in dialogue research that timing is an important variable in interaction (Bouwhuis, 1993). Hence, long and variable response times can have unpredictable and certainly non-intended effects on the student-system interaction. One may also observe that the complexity of the educational system grows as fast as the processing power of the computer platform.

3.2 The Microworld Approach

Nowhere in the educational technology arena has the computer been employed with such striking effects as in the so-called microworld's. Strictly speaking they are not Intelligent Tutoring Systems, but rather more exploratory simulation worlds in which selected phenomena from various scientific fields can be simulated. In a number of instances they can continuously accept input from the student, on the basis of which a process can be continued; sometimes they only show the process evolving from initial conditions set by the student. Popular topics are Newtonian kinematics and chemical titration; an interesting alternative topic is refraction through optic media (Reiner, 1993). Their most exciting property is that they can show at arbitrary time scales and space scales the exact evolution of a particular process that otherwise would be unobservable and difficult to repeat, but also hard to understand and to predict. Such has also been the main motivation behind them. Designers usually stress especially the intrinsic possibility of providing sudden insight into particular well-defined phenomena. The correction of a persistent misunderstanding is indeed one of the most striking occurrences in microworlds.

3.2.1 Whatever Happened to Spacelab?

Spring 1993, BBC's scientific programme Horizon gave an illuminating example of the difficulties encountered in trying to build Spacelab. The original idea was to launch heavy modules making up the lab into orbit, where, being weightless, they could be easily assembled by astronauts without any need for cranes and lifting equipment, that, because of their weight, would have reduced the useful payload to an extent that the mission would become infeasible. Once in space the modules were indeed weightless but, unfortunately, had not lost an ounce of their mass and so retained the same amount of inertia as on earth. Astronauts were almost unable to manipulate the sheer bulk, and risked being crushed between the lumbering masses. As a result the Spacelab mission was seriously delayed and strongly handicapped by the need to provide heavy and energy consuming moving equipment. How was it possible that the eminent physicists behind the plan overlooked this simple phenomenon and confused weight with mass? The example demonstrates how strongly humans are influenced by their everyday experience and perception, despite their objective and generalizable knowledge of the physical world. Sadly, most people get around quite successfully in their environment, coping with movable objects on sliding surfaces, with largely erroneous beliefs about mass, force and friction. Their measure of success, then, is not so much determined by the intrinsic knowledge of process properties but by the cumulative experience of the skilled operator, sportsman, production worker, or any inhabitant of this earth. Erroneous beliefs about the world may never be invalidated in the subset of the physical world people are confronted with, and so still be of great use to their human adherents. The Spacelab example shows that it is futile to try to completely stamp out incorrect subjective models of the world. To put education at fault seems really too easy, where, in fact, human nature is not sufficiently well understood. Yet it is examples like this that are most frequently employed to foster the microworld approach that should attain this goal. The learning goal of the microworld system is therefore disputable. Though the microworld may be quite productive to the professional in discovering unknown phenomena, it is also implicitly assumed that this discovery will lead to durable and deep understanding by naive students. There is in fact no reliable and systematic evidence on this; the incidental nature of the discovery seems at most to lead to fragmentary knowledge, not to a generalizable body of knowledge. It may equally well lead to confusion and increased cognitive load, as the simulation is, by definition, not under user control and does not synchronize its progress with the student's understanding. An additional problem is the demand character. For some reason, microworlds seem to be aimed at group teaching, and from the existing protocols it emerges that frequently only few of the participants are actively engaged in solving the simulation problem. Trial and error seem often to be employed rather than insight in order to make the simulation eventually "work". The nature of graphical presentation may be critical too; di Sessa (1994) relates that some children in the BOXER microworld thought that more closely spaced dots, indicating the course of a falling object, suggested a higher speed, whereas increasing speed actually makes isochronous dots thin out. As falling objects in our world rarely leave a trail of dots, the veridicality of the presentation in microworlds may be more crucial than many theorists think.

3.3 Computer-Based Training: Drill-and-Practice

The need for remedial teaching, either due to insufficient teaching or to learning difficulties, still motivates the production of a staggering amount of computer based drill and practice programs. Few of them find their way to the scientific community, but some can still be found in the NATO Series, always featuring a humble apology. Far more intelligent systems are supposed to have superseded them, so what is it that they have to offer? First, they are generally easy to design, restricted as they are to well-defined domains. But they also train skill acquisition, one aspect that ITSs do not deal with, but which is an important component of learning. Control procedures in drill and practice programs are usually extremely simple, and certainly lack the sophistication of ITSs in probing the knowledge state. Whereas an ITS would stop testing, a CAI program would continue with more practice, that may be logically unmotivated, but hones the expertise anyway. While the drill and practice program aims at understanding, it hardly, or not at all tests for it in a real sense, and so its repetitive character generally reduces its motivating value. Clark and Craig (1992) observe that learning systems initially cause large learning gains owing to being perceived as novel, a phenomenon sometimes referred to as the Hawthorne effect (Gillespie, 1991). As in regular and school-based use the novelty and the challenge wear off rather soon, drill and practice programs have a particularly low survival rate, even if they are attractively designed and provide a good deal of inter action. So, their existence, in turn, motivates the continued interest in (potentially) more challenging systems, like ITSs. An surprising and worrying finding in relation to the classroom use of CBT is teacher involvement. Teachers, when accepting CBT in their class, stipulate that they can include their own material in the program, that they want to control the assignment of exercises to individual children, and want to command an administration in order to monitor pupil progress. In actual practice no such thing happens. In classroom trials of the Reading Board (Ellermann, 1991) where teachers wishes had been carefully implemented in a participatory design procedure, none of five participating teachers ever so much as touched the system as long as it was in the classroom. In a much wider 1988 survey in the South of the Netherlands on the usage of a reading program by an educational publisher it turned out, rather unexpectedly, that none of the 200 teachers in the study intervened in the children's use of the program. All children effectively controlled the activities themselves; for the better or the worse. A connected issue is individual participation. It turns out that as soon as system use is not formally scheduled, individual participation rate becomes strongly biased towards those children who are good learners at the expense of the poor learners. Consequently, one of the main aims of CBT to support especially the poor performers, made possible by the non-frustrating character of the computer, may be foiled by the intricate social relations that pervade the innocuous classroom.

3.4 Distance Learning

In contrast to most systems discussed above, there are learning systems that can almost only be developed in actual practice (for an ambitious exception see Smith, 1992). As a rule these are networked systems, essentially built on a conventional electronic mail system in which pupils interact by means of a very simple send/receive protocol on solving, e.g. arithmetical problems. CSILE by Bereiter and Scardamalia (1992) is a case in point; Romiszowski & Chang's (1992) Computer-Mediated Communication (CMC) another. Further examples are provided by Newman (1992) describing the Earthlab program, and by Vossen and Hofmann (1992) who implemented a collaborative learning and constructing environment. What most of these studies show is that, provided adequate and relatively intensive monitoring and prompting is applied by a supervising teacher, notable improvements are obtained in solving various cases and language assignments by children that in a classroom context score low, or fail. Romiszowski and Chang (1992) note that, in contrast to traditional CAI, the level of student participation in their CMC increased for prolonged periods. In addition, the quantity and quality of the contribution of foreign students improved markedly, compared to that what was attained in a regular classroom setting with delivery teaching. The total amount of student messages at the start was about half that of the teacher, but at the end of the semester it had increased to nine times as much, whereas the volume of teacher output remained fairly constant. Obviously the intense communication within the student network strengthens the demand character far beyond the level obtainable by other systems. Not all students, however, liked the system equally well; there was a 2 to 1 preference for distance learning, suggesting a lower appreciation than the practically unanimous endorsement with which novel CBT is embraced. The intriguing thing to note is that the supporting system, a computer network, has no student or tutoring models at all, but apparently provides the basis on which successful learning can take place. The system itself does not teach, it does not even have a learning action interface; it only sub serves communication. Note too that the teacher is back again. The context of learning is the continuous but virtual presence of other pupils, sending messages, asking for help, comments, gossip and encouragement; and a teacher who eventually will contact and correct even the most secluded and taciturn pupil. Whether the total learning gains obtained with distance learning systems will outstrip conventional teaching remains to be seen, but the relatively low investment, and the selective benefit for poor learners warrants further evaluation.

4 Discussion

4.1 Economics of Instruction

One might argue that the use of rule-based teaching systems, e.g. ITSs, is logically defensible by the formal efficiency of teaching that it entails in representing vast knowledge structures in a highly economical and productive way. However, obtaining learning goals with tutoring systems begs the question of economic feasibility. The main question is then whether the high cost spent in lengthy development and heavy investments will ever be recouped, especially in situations where learning results are erratic, sometimes negative, limited to very restricted domains, and traditional methods yield at least comparable and more stable results. Most ITS projects, in designing their intelligent tutors, even ignore the existing knowledge base of instructional theory (Reigeluth, 1992). Adaptive ITSs are not adaptive in the sense that they will change the architecture of the conceptual domain and the teaching strategies as a result of their application in actual education; there is not a feedback loop with a sufficiently wide bandwidth to relate the effects of training to the system's operation. It is a telling sign that practically all discussions on ITSs contain a section starting with something like: "We have still many years to go before..."; "...and maybe we will never reach that goal." (Psotka, Holland & Kerst, 1992); "We need a set of entirely new principles for ITSs"; "We realize just how daunting the obstacles are!" (Costa, 1992; McCalla, 1992); "The knowledge of learning results is too immature to be incorporated in ITS's (Merrill, Li & Jones, 1992). In the broader framework of Educational Systems at large, the situation is not much different. The most comprehensive volume on educational systems design (Reigeluth, Banathy & Olson, 1993) in the NATO ASI series treats the issue of evaluation of learning performance only in passing. Reigeluth (1993) mentions 15 different activities under which the principles of educational system design have to be nested; not one of them is concerned with evaluation of the system, either in the laboratory or in actual practice. Others flatly deny the ability of "traditional econometric/statistical methods" to reveal system performance (a.o. Banathy, 1993). It is worth mentioning that in less ambitious and strongly domain-focused approaches (traditional) data have been obtained, that show the specific benefits of an interactive learning system impressively clearly.

4.2 Learning and its Evaluation

In scrutinizing the extant literature one is confronted with one of the most baffling circumstances in the field. In the 25+ NATO workshop volumes that were available at the time of preparing this overview we have not encountered a single system that was employed on a permanent basis in actual education in schools or universities. Despite the past decades of development of instructional systems, not a single system of these has ever been put to successful and permanent use in realistic educational settings (Allen, 1993), except for relatively small student populations. Designs and implementations of computer-based training systems, in spite of massive funding and widespread investigations and recipient of an almost exclusive attention by instructional scientists, show embarrassingly clearly that the emphasis is on the use of the computer, not on the benefits for the learner. This means that there is no evidence how effective these systems will be in formal education, and to what extent they contribute to actual learning of the kind the system designers have in mind. Exceptions to this general statement are employee training in enterprises and, probably, operator training for military systems. Gayeski (1992) provides some data on CBT in companies, and observes that most training, as stipulated in the introduction, is involved with learning basic computer skills: word processing, spreadsheets and accounting. Even for this modest application more than 50% of all companies who employed interactive video projects for training abandoned it. Those CBT programs that were retained were mostly made by internal staff, were in linear form, and abolished all interactive video, multimedia and expensive platforms. Gayeski (1992) observes that only participatory design of training programs inside the company has ultimately been successful. Data on learning by educational technology systems in public education are conspicuously scarce, but when available they do not constitute reassuring reading. The most comprehensive overview, covering an impressive number of educational technology comparison studies by Clark and Craig (1992) shows that learning gains, if any, are only observed in the initial period when systems are considered to be novel; after that learning performance gradually converges to the same performance as attained with conventional methods. This same conclusion was reached by other investigators as well (e.g. Hagler & Knowlton, 1987) and even understandable reactions (Kulik, Kulik & Bangert-Drowns, 1985) are unable to change those findings. As referred to above, many studies on guided discovery learning, usually in scientific domains, report actually lower learning scores obtained with the specially designed educational methods (Bork, 1992, McCalla, 1992). While many educational researchers promote strongly the idea of multimedia technology, Clark & Craig show convincingly that the multimedia interface is often detrimental to learning and that the supposed advantage of these media reflects largely incorrect ideas concerning perceptual processing and acquisition. A similar appeal as for multimedia is shared by Hypertext and here again the empirical evidence concerning its effectively is disappointing (Hammond, 1992; Thimbleby, 1992; Draper, 1992).

4.3 Successful Teaching or Learning?

There are, however, spectacularly successful examples of computer-based training, that are conspicuously absent in the entire range of NATO workshops. The most important one is the flight simulator, that in its field has conquered more than 50% of total training time, and which models a tremendously complex system, but still, basically, only trains skills. The premises of the flight simulator are validated in actual pilot's performance, and operating it is like interacting with the real thing. Again, there is no student model or a tutoring model in the flight simulator; the trainee pilot and the simulator act and react in response to one another as equal partners. The same holds for the much less expensive and less perilous video arcade games, where there is not even a real thing, but interacting with it is eagerly learned. Both systems feature a very intense, sometimes addictive interaction dialogue, but do not contain any student model, nor conceptual domain knowledge. How can they possibly work? The potential objection that, e.g., language learning is a far more abstract and complex activity than operating a slot machine is easily refuted by the fact that little children develop remarkable linguistic ability at an age at which none of these demanding control tasks would be entrusted to them. Their parents do not employ linguistic expert systems or elaborate error interpretation schemes; they recklessly stimulate any understandable speech utterance, often do not bother to correct and hardly know a single grammar rule. It is an appealing question to analyze what constitutes the challenging character of the environments in which learning is pursued so effectively and passionately. Despite the ignominious character of arcade games, there is actually some theory on game-playing behavior backed up by systematic empirical research that is well-known by game designers (Skelly & Thiel, 1990; Czikszentmihalyi & Czikszentmihalyi, 1988)

4.4 The Need for Cognitive Research

The instruction-learning dichotomy presupposes that there is by default reception, learning and memory at the learner side, complementary to instruction. This second part is systematically neglected in educational system development. Learner variables are in fact rarely taken into account, e.g. learning ability, developmental/genetic issues, memory organization and load and the nature of cognitive representation, At most, training systems may be intended for a group with specific learning disabilities, e.g. dyslexia., but tutoring systems do not address the issue of differences in individual learning ability. The volume edited by Ferguson (1993) is interesting in this respect as several of the chapters report investigations of the cognitive representations of physical concepts by students. They show that cognitive representations may vary so widely, that a single educational system must be considered to be incapable of capturing the variation. There is also clear evidence that educational technology affects students differently; for some it works effectively, while it fails for others, and the same holds for teachers. Maturation, too, may play a determining role in the nature of interaction and the kind of errors made by pupils, which may ruin the diagnostic validity of tutoring systems (e.g. the genetic variable discovered by Beishuizen and Felix, 1992). As argued before, cognitive representation is treated ordinarily in the context of the specific architecture of the tutoring system, hardly or not at all as a research issue in its own right. Where it does (MacWhinney, 1992; Miller, 1992), it supplies a fascinating insight in the actual organization of e.g. the lexical representation, and semantic networks. Such models are derived from and evaluated by research on subjects, potential students. Research of this kind should be strongly supported and extended in order to supply valid information concerning human information processing and knowledge acquisition and representation.

4.5 Locus of Control and Agents

Central concepts like "Interactivity" and "Locus of Control" are variously defined and the topic of considerable confusion. The seminal paper of Gentner (1992), however, points the way towards a research field that may be of basic importance to any interactive training system. The interesting aspect of the concept "locus of control" is actually that it can only be understood to its full implications in systematic research on actual student or user interaction in a range of learning and operating tasks. The further elaboration of the concept is intimately related to the development of agent theory. Hard problems in this area are predicted by a number of authors (Baker, Brazdil, Coelho, 1992), but the essence, in view of some of the implementation difficulties as sketched above, seems to be to keep the agents as simple and unintelligent as possible (Connah,1994). This may seem contrary to the almost universal desire to make things maximally intelligent, but what should count is the effectively in the learning domain and the robustness with which it can be obtained. A comparison with the flight simulator in this respect is useful in that it both provides an interesting application of locus of control, but also of the operation of simple agents, natural ones or other. The low esteem of traditional educational methods, a time-honored concept, may also be responsible for the casual negligence of the importance of fitting the new instructional method to educational practice. It would be nice to see that robustness of operation would become a real yardstick for effectively and acceptability.

4.6 The Context of Instruction

While it can be stated that the situationist approach has made valid points concerning the context in which learning takes place, stressing social apprenticeship and community of practice, directly usable results for implementing learning systems have not been produced, while the issue of evaluation is at most anecdotal. In this same context the design and use of "micro-world" learning systems is usually seen as directly combating lowered educational success in many Western countries. No doubt researchers are stimulated by the impressive motivation of children working with the microworld tools (Vivet, 1992), but systematic evaluation and follow-up studies are scarce. Thornton, (1993) is one of the few who carried out extensive comparison studies and obtained very significant effects in favor of his Microcomputer-Based Laboratory tool on Newtonian kinematics and dynamics. Yet, it has to be realized that in comparison studies age levels are frequently different as well as the topics learned and the method of presentation (White, 1992). So, the identification of what exactly is learned in microworlds is particularly hard. Radiant insights obtained with these systems cannot obscure some wishful thinking involved. A sobering thought concerning distance learning is in order too. The initiative and excitement that participants develop in the course of time may produce completely unpredictable effects, that may lead to learning very different things from the intended ones. It may lead to something quite different from learning, but above all, to the realization that many explicit wishes can simply not be satisfied at the same time with the existing infrastructure, and sometimes any infrastructure. Robustness in distance learning is not guaranteed. Claims aside, introduction of new instructional methods should in principle only be motivated either by measured superiority at the same cost, or the same effectively at lower cost. Empirical research on evaluation of this kind is mostly absent, fallacious or focused on too small details. One issue, consistently skirted, is the long-term effectively of instruction methods. Usually the technical life span of an instructional system is far shorter than the productive life span of the (former) learner, which impedes effective feedback towards redesign, but also obscures the cost issue.

5   Conclusions

1. Implementation and evaluation are weak points in AET research. Consequently the beneficial feedback of reality is missing and the value of theories cannot be assessed.
2. Hierarchical system architectures as proposed for regular ITSs are probably too complex for practical usage. Distributed architectures, such as in agent theory, are more promising.

3. Concerted efforts directed at specific interactive learning systems for areas of great social and economic value would offer great prospects. Such projects should carefully be planned and implemented because of the many interest groups involved.
4. Alternatives for ITSs have been proposed, such as Hypertext, microworlds and agents. In the absence of positive evaluation of what is being learned, it is difficult to assess their lasting value. Nevertheless, intense student involvement as with microworlds will be an important aspect of any fruitful interactive learning system and recent agent theories do at least circumvent long standing problems of ITSs.
5. Systematic investigation of the development of mental models in the course of learning provides basic insights in the learning process that should be at the heart of educational design.
6. A structural approach to the design of instructional dialogue and interfaces is urgently needed.

References

(NATO ASI Series F volumes are indicated by F and volume number)

Allen, D.W. (1993) School reform movements: Tinkering with the system. In F95
Anderson, J.R., Boyle, C.F., Corbett, A.T., and Lewis, M.W. (1990). Cognitive modeling and intelligent tutoring. Artificial Intelligence, 42, 7-49
Baker, M.J. (1992) Negotiating goals in intelligent tutoring dialogues. In F91
Banathy, B.H. (1993) Systems design: A creative response to the current educational predicament. In F95
Beishuizen, J. and Felix, E. (1992) Adaptive tutoring of arithmetic skills. In F104
Bereiter, C. and Scardamalia, M. (1992) Two models of classroom learning using a communal database. In F104
Beun, R.J., Reiner, M. and Baker, M. (eds.) (1994) Dialogue and instruction. F162
Bosser, T. (1992). Interactive environments for learning and tutoring. In F87
Bork, A. (1992) Learning in the twenty-first century interactive multimedia technology. In F93
Bouwhuis, D.G. (1993) Speech technology in interactive instruction, In: R. Bennett, A. Syrdal and S. Greenspan (eds.) Behavioral aspects of speech technology: Theory and practice. Amsterdam: Elsevier Science Publishers
Brazdil, P.B. (1992) Integration of knowledge in multi-agent environments. In F91
Brown, J.S., and Burton, R.R. (1978) Diagnostic models for procedural bugs in mathematical skills. Cognitive Science 2, 155-191
Carbonell, J.R. (1970) Mixed initiative man-computer instructional dialogues. MIT: PhD thesis
Chall, J. (1967) Learning to read: the great debate. New York: McGraw-Hill

Chall, J. (1977) The great debate: ten years later, with a modest proposal for reading stages. In: L.B. Resnick and P.A. Weaver (eds.) Theory and practice of early reading, Vol. 1. Hillsdale, NJ: Lawrence Erlbaum Associates

Clancey, W.J. (1987). Knowledge-based tutoring: The GUIDON program. Cambridge, MA: MIT Press

Clark, R.E., and Craig, T.G. (1992) Research on theory on multimedia learning effects. In F93

Coelho, H. (1992) Facing hard problems in multi-agent interactions. In F91

Connah, D.A. (1994) The design of interacting agents. F142

Costa, E. (1992) The present and future of intelligent tutoring systems. In F96

Czikszentmihalyi, M. and Czikszentmihalyi, I.S. (1988) Optimal experience – psychological studies on flow in consciousness. NY: Cambridge University Press

Draper, S.W. (1992) Gloves for the mind. In F81

Ellermann, H.H. (1991) MIR: A monitor for initial reading. Doctoral thesis, Eindhoven University of Technology

Ennals, R. (1992) Computers and exploratory learning in the real world. In F81

Evans, D.A., and Patel, V.L. (eds.) (1992) Advanced models of cognition for medical training and practice. F97

Ferguson, D. (ed.) (1993) Advanced educational technologies for mathematics and science. F107

Frasson, C. (1992). From expert systems to intelligent tutoring systems. In F97

Frederiksen, J.R. and White, B.Y. (1992) Mental models and understanding: A problem for science education. In F96

Gayeski, D.M. (1992) Enhancing the acceptance and cultural validity of interactive multimedia. In F93

Gentner, D.R. (1992) Interfaces for learning: Motivation and the locus of control. In F87

Gentner, D., and Stevens, A. (eds.) (1983) Mental models. Hillsdale, NJ: Lawrence Erlbaum Associates

Gillespie, R. (1991) A history of the Hawthorne experiments. New York: Cambridge University Press

Gluck, M., and Bower, G. (1988a) Evaluating an adaptive network model of human learning. J. of Memory and Language 27, 166-195

Gluck, M., and Bower, G. (1988b) From conditioning to category learning: An adaptive network model. J. of Experimental Psychology: General 177, 227-247

Grandbastien, M. (1993) New educational technologies cannot be fully integrated in existing educational systems. In F95

Hagler, P., and Knowlton, J. (1987) Invalid implicit assumption in CBI comparison research. J. of Computer-Based Instruction 14 , 84-88

Hammond, N. (1992) Tailoring hypertext for the Learner. In F81

Hron, A. (1992) Simultaneous processing of different problem aspects in expert problem solving: An analysis in the domain of physics on the basis of formal theories of common sense knowledge. In F86

Hron, A., Bollwahn, J., Mandl, H., Oestermeier, U., and Tergan, S.-O. (1992) Computer-based learning environment and automatic diagnostic system for superimposition of motion. In F86

Hunt, E. (1989) Connectionist and rule-based representations of expert knowledge. Behavior Research Methods, Instruments and Computers 21, 88-95

Jones, M., and Winne, P.H. (eds.) (1992) Adaptive learning environments. F85

Shuell, T.J. (1992) Designing instructional computing systems for meaningful learning. In F85

Kaye, A.R. (ed.) (1992) Collaborative learning through computer conferencing. F90

Kulik, J.A., Kulik, C-L.C. and Bangert-Drowns, R.L. (1985) The importance of outcome studies: A reply to Clark. J. of Educational Computing Research 1, 31-387

Lundell, J.W. (1988) Knowledge extraction and the modeling of expertise in a diagnostic task. Doctoral dissertation, University of Washington, Seattle, WA

MacWhinney, B. (1992) The competition model and foreign language acquisition. In F80

Mandl, H., and Hron, A. (1992) Cognitive theories as a basis for student modeling.In F86

Mathews, M.M. (1966) Teaching to read. Chicago: University of Chicago Press

McArthur, D., Statz, C., Hotta, J., Peter, O. and Burdorf, C. (1988) Skill-oriented task sequencing in an intelligent tutor for basic algebra. Instructional Science 17, 281-307

McCalla, G.I. (1992) Intelligent tutoring systems: Navigating the rocky road to success. In F96

McDermott, L.C. (1992) Research as a guide for the design of intelligent learning environments. In F86

McSpadden, M. (1989) Remembering the future. Unpublished manuscript

Merril, M.D., Li, Z. and Jones, M.K. (1992) An introduction to instructional transaction theory. In F104

Newman, D. (1992) Formative experiments on the coevolution of technology and the educational environment. In F96

Norman, D. (1981) Categorization of action slips. Psychological Review 88, 1-15

Norman, D. (1993) Cognition in the head and in the world: An introduction to the special issue on situated action. Cognitive Science 17, 1-6

Norman, D., and Draper, S. (eds.) (1986) User-centered system design. Hillsdale, NJ: Lawrence Erlbaum Associates

Nosofsky, R.M., Clark, S.E., and Shin, H.J. (1989) Rules and exemplars in categorization, identification, and recognition. J. of Experimental Psychology: Learning, Memory and Cognition 15, 282-304

Nosofsky, R.M., Kruschke, J.K., and McKinley, S. (1992) Combining exemplar-based category representations and connectionist learning rules. J. of Experimental Psychology: Learning, Memory, and Cognition 18, 211-233

O'Shea, T. (1979) Self-improving teaching systems. Ph.D. thesis, University of Leeds

Payne, S., and Squibb, H. (1990). Algebra mal-rules and cognitive accounts of error. Cognitive Science 14, 445-481

Ploetzner, R., and Spada, H. (1992) Analysis-based learning on multiple levels of mental representation. In F84

Psotka, J., Holland, M., and Kerst, S. (1992) The technological promise of second language intelligent tutoring systems in the 21st century. In F80

Reigeluth, C.M. (1992) New directions for educational technology. In F96

Reigeluth, C.M. (1993) Principles of educational system Design. In F95

Reigeluth, C.M., Banathy, B.H. and Olson, J.R. (eds.) (1993) Comprehensive systems design: A new educational technology. F95

Reimann, P. (1992) Eliciting hypothesis-driven learning in a computer-based discovery environment. In F86

Romiszowski, A. and Chang, E. (1992) Hypertext's contribution to computer-mediated communication: In search of an instructional mode l. In F93

Roschelle, J. (1990) Designing for conversations. Paper presented at the Annual Meeting of the American Educational Research Association, Boston

Rypa, M. (1992) CALLE: A computer-assisted language learning environment. In F87

Self, J. (1974) Student models in computer-aided instruction. Int. J. of Man-Machine Studies 6, 261-276

di Sessa, A. (1994) Designing Newton's laws: patterns of social and representational feedback in a learning task. In F142

Skelly, T.C., and Thiel, D.D. (1990) The development of seductive interfaces. Tutorial Notes, CHI'90, Seattle, WA

Smith, R.B. (1992) A prototype futuristic technology for distance education. In F96

Snow, R.E., and Swanson, J. (1992) Instructional psychology. Annual Review of Psychology 43, 583-626

Soloway, E., Guzdial, M., Brade, K., Hohmann, L., Tabak, I., Weingrad, P., and Blumenfeld, P. (1992) Technological support for the learning and doing of design. In F85

Spaai, G.W.G., and Ellermann, H.H. (1990) Learning to read with the help of speech feedback: An evaluation of computerized reading exercises for initial readers. In: J.M. Pieters, P.R.J. Simons and L. de Leeuw (eds.) Research on computer-based instruction. Lisse: Swets & Zeitlinger

Stucky, S. (1992) Situated cognition: A strong hypothesis. In F87

Taylor, M.M., Neel, F., and Bouwhuis, D.G. (eds.) (1989) The structure of multimodal dialogue. Amsterdam: North-Holland

Tiberghien, A., and Mandl, H. (eds.) (1992) Intelligent learning environments and knowledge acquisition in physics. F86

Tinker, R.F., and Thornton, R.K. (1992) Constructing student knowledge in science. In F96

Thimbleby, H. (1992) Heuristics for cognitive tools. In F81

Thornton, R.K. (1993) Enhancing and evaluating students' learning of motion concepts. In F86

Vivet, M. (1992) Research in advanced educational technology: Two methods. In F96

Vosniadou, S. (1992) Modelling the learner: Lessons from the study of knowledge reorganization in astronomy. In F86

Vossen, P., and Hofmann, J. (1992) Using Timbuktu™ and Guide™ for Computer-Supported Group Learning. In F81

Wenger, E. (1987) Artificial intelligence and tutoring systems. Los Altos, CA: Morgan Kaufman

White, B.Y. (1992) A Microworld-based approach to science education. In F96

Wielinga, B., Schreiber, T., and Breuker, J. (1992) KADS: A modeling approach to knowledge engineering. Knowledge Acquisition 4, 5-53

Winkels, R. (1992) Modelling skill learning. In F87

Witschital, P. (1992) Why we need "composable" user interfaces. In F87

d'Ydewalle, G. (1992) Knowledge acquisition and knowledge representation. In F87

Zucchermaglio, C. (1992) Toward a cognitive ergonomics of educational technology. In F105

3

Collaborative Distance Learning and Computer Conferencing

France Henri, Claude Ricciardi Rigault

Abstract: Computer conferencing used in pedagogical contexts is proving to be a gold mine of information concerning learning. Educators, trainers and learners should be able to use this information to enhance the learning process. A research project is presently under way to develop intelligent tools to analyze the content of messages. For educators and trainers, a tool will help to decrypt the learning process in order to give appropriate support to learners. For learners, a series of tools will assist construction and formalization of knowledge, elicit awareness of their socio-cognitive processes and stimulate improvement of the quality of their participation.

Keywords: Computer conferencing, learning process, cognitive process, metacognitive process, social dynamics, interaction, group learning, collaborative learning, cooperative learning, message analysis, distance learning, distance education

1 Introduction

Until very recently, the use of traditional non-interactive media in distance education only provided trainers with the option of pedagogic methods in which individual learning is predominant. Learners in distant geographical locations were left isolated and, thus, deprived of learning methods that originate from interactive communication and a social dynamic. Henceforth, however, through virtual environments, learners can be brought together, thanks to communication interactive technologies, for the purpose of communicating among themselves. It is now possible to apply, from a distance, the kinds of pedagogic strategies that make use of the social and cognitive contexts of group interactions. This possibility, novel as it is, is another milestone in distance education pedagogy which has often been criticized for using such an uninteractive and too transmissive an approach. (Henri and Kaye, 1985; Jarvis, 1986; Evans and Nation, 1989; Jacquinot, 1990).

Collaborative learning, generally considered to be a method reserved exclusively for face-to-face situations, is among the new pedagogic group approaches that distance education, in this new technological age, is gradually adopting. Several experiments and studies have shown that computer conferencing is particularly suitable in sustaining a distance collaborative learning process for adult students. It is from this perspective that our arguments as well as our research orientation in this area will be examined.

First, we shall attempt to describe collaborative learning and cooperative learning. These are two very similar approaches which are often confused, but must be distinguished from each other when designing learning activities through teleconferencing or when implementing and, especially evaluating them. Secondly, we shall examine the characteristics of teleconferencing and the way it is used in distance education in order to demonstrate what, on the one hand, enables it to sustain the collaborative learning process and, what on the other, hinders its application. This study makes it possible to develop a specific mode of organization and a set of implementation principles that are different from those that obtain in face-to-face situations. Lastly, we shall examine the progress we have made so far in our research which aims essentially at developing appropriate tools for teleconferencing tutors and moderators to help and support groups during a collaborative learning process.

2   Collaborative Learning and Cooperative Learning

In citing work related to collaborative and cooperative learning, writers have too often referred to collaborative learning and cooperative learning interchangeably and without distinction. But when we try to transpose these processes (that are generally poorly defined in technological contexts), when we are faced with the problem of applying "theories" in real and practical situations, we quickly realize that there are nuances that call for a distinction between collaborative group learning and cooperative group learning. A rather byzantine or trifling distinction for some, but extremely useful and fundamental for those who must work out the mechanics of the processes and ensure coherence between ideas that underlie theory and those that underlie practice.

2.1 Collaboration and Cooperation: Structured Processes

There is a clear difference between group learning and collaborative or cooperative learning. Group learning is that which originates from simply bringing learners together for discussion, exchange, interaction and mutual assistance. But to qualify as cooperative or collaborative, the process must be structured in the following ways:

- members should work and interact in a strictly interdependent manner;
- they should be committed to encouraging each and everyone to learn;
- they should feel individually duty-bound to discharge the functions assigned to them;
- they should use the appropriate social and interpersonal skills to stimulate cooperation with a view to achieving success or attaining the set goals; and, lastly,
- they should revise the group-functioning process.
(Johnson, Johnson and Smith, 1991)

Interdependence, which is a central asset of the approach and the driving principle of the process, must be applied positively within the framework of equality and mutual respect in small groups (Hooper, 1992).

Writers suggest that cooperative or collaborative learning in face-to-face situations is not meant for large groups; in fact, it could be restricted to five or six persons at most. The approach focuses on heterogenous groupings where learners can meet other persons with diverse aptitudes and skills so that they can all pull their respective resources with a view to attaining the set goals in a spirit of positive interdependence.

Hooper (1992) proposes that to distinguish the collaborative and cooperative approaches from other approaches, the nature of the interactions should be compared on the basis of the following criteria; the interactions should:

- facilitate mutual assistance and confidence;
- be a source of encouragement;
- contribute to the creation of a relaxed and calm atmosphere;
- encourage the sharing of skills;
- aim at an efficient and effective processing of information;
- tend to criticize the proposals and opinions expressed with a view to achieving greater quality of work and to enriching decisions.
(Johnson, Johnson and Smith, 1991)

In the implementation of the collaborative and cooperative process, the teacher plays a fundamental role as facilitator. He has the delicate task of forming groups, structuring the activity, supporting the group in its work and, especially of ensuring that a revision of the group functioning process takes place. He must help the members to specify to what extent group actions have contributed to the accomplishment of the task and to identify conducts to be maintained, improved upon or changed. Through these interventions, he helps the group to learn to judge itself. The purpose of this approach is to:

- enable the students to concentrate their efforts on the maintenance of good working relations among the members;
- facilitate the learning of cooperation skills;
- guarantee that members receive a feedback of their participation;
- reinforce the positive conducts identified among group members;

- ensure that students maintain a cognitive as well as metacognitive frame of mind and, lastly,
- provide a means of ascertaining the success of a group.
  (Johnson, Johnson and Smith, 1991)

The teacher's task, therefore, includes observing groups, making an analysis of group-functioning problems and providing a feedback on group-work methods.
   This approach offers a lot of advantages to the student. They include:

- the development of greater cognitive skills;
- an increase in the level of understanding (acquisition);
- the growth of satisfaction in relation to learning experience;
- the development of positive attitudes towards the subject studied;
- the improvement of communication skills as well as social and interpersonal skills;
- the strengthening of self-esteem and,
- the accommodation of cultural and racial differences.
  (Cooper et al., 1990)

   Much as the collaborative or cooperative processes aim at acquiring knowledge, they also aim at contributing to the development of the 'social' aspect of the human being. Thus, the group must observe the exact working conditions, and its activity must be structured with a view to achieving the common objective.

2.2 Collaboration and Cooperation Distinguished

According to a study conducted at Concordia University by the Center for the Study of Classroom Processes (Abrami, 1993), collaborative learning is not just a variant of the cooperative learning method; rather, it is an approach rooted in theories and philosophies propounded by certain sociologists, like Karabel, who is concerned about the fact that the school is an hierarchical institution and that it reflects and perpetuates the status quo in society. Advocates of collaborative learning seek to transform society by restructuring power relations within the school into a more egalitarian framework, while at the same time aiming at developing the social aspect of the human being as in the cooperative approach (Pradl, 1991). They call for a more democratic process, one that opts for an equal distribution of power among learners and within the school structure.
   The study further reveals that the collaborative approach adopts a less structured method which is very much based on the intrinsic interest of the learners in accomplishing their tasks. Equality prevails not only among the group of learners, but also between the teachers and learners; and a spirit of openness and shared responsibility reigns among them. In this way, learners enjoy greater autonomy in the collaborative approach than in any cooperative approach. They can choose those with whom they want to work rather than allow the teacher to form heterogeneous groups; they can hold discussions with the teacher on what should

be learnt and the learning method to be adopted, rather than leave him take these decisions alone (Boomer, 1990).

Hooper (1992) for his part, believes that the distinction between collaboration and cooperation is based on the structure of the task. He states that with the cooperative method, the accomplishment of tasks can be achieved either through collaborative effort or through task-specialization. The collaborative task calls for the parallel participation of each group member. If, for instance, fifty words of a foreign language are to be learnt, each group member will have to learn the fifty words and provide learning assistance and support to the other members to learn them. Task-specialization, on the contrary, will require that each member learn, individually, a sub-set of the fifty words and then teach the others. In this example, task-specialization, therefore, requires members to individually learn part of the lesson and teach it to others, whereas the collaborative method encourages students to mutually assist one another in their efforts to learn the entire lesson. This distinction makes us conclude that the collaborative approach results in reducing interdependence within the groups, and thus, in underscoring the importance of mutuality in accomplishing the task.

In brief, the collaborative approach differs from the cooperative approach in three aspects. First, in autonomy, since the students are free to choose those with whom they want to work, to influence the selection of the lesson to be learnt and to defend their learning methods. Secondly, in the method of task accomplishment, since tasks are not shared or split up among group members. Lastly, in the interdependence of group members, as group functioning does not necessarily guarantee the complementarity of individual skills. The collaborative approach structure is, therefore, more flexible and, consequently, requires less planning and coordination in achieving the objectives.

2.3 Collaborative Learning in Distance Education for Adult Learners

As has been seen already, the collaborative learning process offers at least two major advantages in distance education for adult learners. It enhances learner autonomy and encourages more flexible group functioning. By focusing on student participation in lesson orientation as well as in the choice of the learning method, the collaborative process adopts an entirely andragogic perspective and covers paradigmatic values, like autonomy and learner responsibility – values which are adopted by many distance education institutions (Henri and Lamy, 1989; Henri, 1993). Furthermore, a collaborative approach, compared to the cooperative method, has fewer constraints relating to the organization and functioning of a group. This makes it more acceptable to the distant learner and easier to apply, manage and administer for a distance education institution.

Groupings are rather rare in distance education because of the distance and geographical dispersion of learners. This is also because individual learning corresponds to the model that is traditionally applied in educational systems (Kaye,

1992), and also because it offers very few constraints for the adult learner. Studies conducted in Tele-university (Laurent, 1987; Hotte, 1990), for instance, indicate that several adult learners would prefer distance education because it enables them to go through it individually, at their own pace, without having to grapple with the attendant constraints of any group process or of their personal, family and professional life. The cooperative learning process might, in striving to achieve very interdependent group functioning, worsen the distant learners' constraints, increase their needs for supporting their process and compelling the distance education institution to assist these groups adequately. The collaborative approach for its part, seems to be more flexible and meets the requirements of distance education for adult learners.

3   Computer Conferencing: A Technology Which Promotes Distance Collaborative Learning

Computer conferencing is the computerized equivalent of a meeting: it brings together, in a virtual environment, a group of persons who do not have to travel, to that effect; it facilitates rapid communication in a synchronous or asynchronous manner, as well as the exchange and sharing of textual messages, while maintaining a common trace of exchange (Paquette, Bergeron, Bourdeau, 1992). As regards asynchronous conferencing, in particular, the user can write messages that the computer stores in memory and which other users can read at their convenience. Some conferencing software are equipped with features used for the processing of information contained in the messages such as menus, directories and search functions, etc. Computer conferencing is a device which provides a framework for group collaboration from a distance and which, in pedagogy, can enhance collaborative learning.

3.1   Computer Conferencing and Collaboration

Though very different from face-to-face interactive communication, the dynamic of communication by computer conferencing involves, in some respect, a more intense interaction (Johansen et al., 1979). In computer conferencing, where exchanges are exclusively textual, the decrease in social pressures, which results from the physical absence of people, encourages a greater freedom of expression and more spontaneity. In the case of asynchronous computer conferencing, the absence of people and the asynchronousness of discussion gives rise to a participation dynamic where the struggle for the right of audience never arises. Participants react more to the content of the message rather than to the attributes of the author of the message (age, physical appearance, status) (Hiltz and Turoff, 1982). They can control the time to give their answers and take the time necessary to analyze the message so as to react more effectively. Designed to facilitate the complex process of human communication, computer conferencing has rapidly proven its effectiveness in the taking of decisions and solving of problems in groups (Hiltz and Turoff, 1982). Today, we know that under good conditions, a computer conference, coordinated by a competent and an alert moderator, facilitates the development of cohesion within the group, stimulates a productive interactive dynamic and gives rise to a sort of collective intelligence (Kerr and Hiltz,1982; Hiltz, 1985b).

3.2 Computer Conferencing and Distance Education

Through its communication characteristics, computer conferencing is trying to overcome the shortcomings of traditional distance pedagogy that is basically suited for self-tuition processes and, thus, offers only very few facilities (excluding the telephone) for exchange among learners. Enriched by computer conferencing, distance education facilitates direct contact among learners, as well as between learners and tutors, thereby eliminating isolation and stimulating participation. Interactivity, the rapidity and flexibility of communication, on-going communication with the group, the absence of constraints in time and space, at least, for asynchronous computer conferencing, and the elimination of isolation are all aspects that transform the distance education process.

Many studies[1] show that computer conferencing technology contributes to the enrichment of the distance education process by increasing not just collaboration but its quality as well. An evaluation of learning effectiveness also shows tangible results, even if they vary with circumstances (Hiltz, 1990). Interactivity generates very positive effects on learning and facilitates the development of more stimulating learning activities which enhance participation (Kaye, 1987; Harasim, 1990). Asynchronous, computer conferencing stimulates a more profound reflection on the content. Ricciardi Rigault and Henri (1989) observe that this technology greatly facilitates, in particular, the examination of concepts through *disputatio* (disputation), and that for certain forms of learning, dialogue and interaction are key elements. The possibility to disagree, ask questions and solve problems in a group are the main factors that enrich this technology. The intellectual work group develops its own energy and yields better results than those of an average group member (Hiltz and Turoff, 1982). When problems are solved collectively, various possible solutions emerge from learners, thereby enriching the content and dynamic of learning (Henri and Lescop, 1987).

Through its interactive nature, computer conferencing makes it possible for users to learn in a collaborative manner, and in this way, enhances collective knowledge (Harasim and Wolfe, 1988). The learning content is molded through a collective process in which the knowledge and experience of all the learners are

---

[1] To name just a few: Hiltz, 1985a; Hailes and Richards, 1984; MaCreary and Van Dureen, 1987; Henri and Lescop, 1987; Shapiro et al.,1987 MaCreary, 1989; Mason and Kaye, 1989; Harasim, 1990; Hiltz, 1990; Kaye, 1992; Henri, 1992.

contributory factors (Shapiro et al, 1987). The learner plays an active participatory role, and is very involved in the creation of knowledge that emerges from active dialogue (Meunier and Henri, 1987), the development of new ideas, as well as development of concepts following group discussion of messages.

In addition to facilitating collaborative learning, computer conferencing characteristics generate other advantages relating to the cognitive and psycho social development of the user (Kerr and Hiltz, 1982; Henri, 1992c). At present, communication by strictly textual messages is an effective means, rather than a disadvantage, that facilitates the development of formal thinking in the user (McCreary, 1989). Writing, more than any other form of expression, compels the user to rigorously organize his thoughts, translate same into a coherent message and communicate it in a simplified, authentic and sober manner for easy understanding by all. In this respect, we have already shown (Henri, 1992a) that the way computer conferencing functions urges the user to apply not only the cognitive skills of information processing in structuring and composing messages, but also metacognitive skills in managing their learning process, objectivation skills in contextualizing their interventions, and psychosocial skills in participating in group interactions.

4   A Technology Which Forces Process Transposition

Most of the studies mentioned above adopted an empirical approach. Through close observation of groups in practical learning situations, research tries to assess the quality of computer conferencing by comparing the way it functions to that of a similar face-to-face situation. This approach enables the researchers to identify, what is done better using the conferencing technology, or what is done poorly. Such a goal is pursued as if we were forgetting that research on media had clearly shown that using media makes us do things differently, often with other objectives, and by using evaluation standards of a different kind (Jacquinot, 1993; see also Gavriel Salomon's work).

In comparing the functioning of communication in computer conferencing with what is observed in face-to-face situations, we run the risk of limiting ourselves to a descriptive and an elementary exercise, at the expense of analysis (Mason, 1992). We could therefore loose the essence of our work and forget it is not really the new technology that should be studied, but rather the challenge of implementing a known activity using a different means. Our focus, in wanting to facilitate collaborative learning through computer conferencing is on the implementation of innovation. Essentially, our aim is not to duplicate old models, but rather to create new ones by exploring processes, by using computer conferencing attributes to intensify some of its aspects, and by bringing to light what is not evident in the face-to-face situation that we claim to know. To ensure mastery of the collaborative process, we must progressively develop interaction procedures, different from those that prevail in face-to-face situations.

The most obvious source of data available for such research is, without contradiction, the transcription of existing computer conferences. Ironically, as reported by Mason (1992), it is the one which is least used. So far, only few analyses have been made of this precious artifacts. Some of the results of these analyses are available to us, and they provide information on the categories of messages communicated by participants, the profile of certain participants, the nature of interaction and on the generic categories of computer conferencing.

4.1 The Transposed Process

To implement collaborative learning in a face-to-face situation, and guide the learners in carrying out their activities which, very often, take the form of projects, Reid, Forrestal et al. (1984) propose a process of five stages:

i)   the contribution from the experience and intrinsic knowledge of each learner;
ii) the examination of the subject chosen, using various means, such as: lectures, discussions and observation;
iii) the restructuring of the information collected;
iv) the presentation of the project and what has been taught to a willing audience, and
v)   a feedback on what has been learnt and on the learning process.

Easily applicable in a face-to-face situation, this process is extremely complex when it has to be applied in a distance education context. The several stages that are involved in the process presuppose that several types of discussion are planned and held, and that the many learner initiatives, as well as the follow-up of individual and collective work are coordinated. The process, therefore, implicitly creates a heavy management burden, which determines the success of a collaborative method.

To attenuate this difficulty, the application of such a process, from a distance, requires, first of all, *a priori* consideration of the management function and the availability of a specific forum for the exercise. Furthermore, the number of stages should be reduced, without at the same time, compromising the process. Of the five stages, two stand out distinctly: the exploration, collection and restructuring of information (steps 1, 2 and 3), and the presentation of a project designed by using information and the feed-back on the activity (steps 4 and 5). Reduced in this way, the stages are easily applicable from a distance. Such a reduction does not affect the quality of the learning process because it takes into account the cognitive strategies of knowledge acquisition and learners' needs. To support our argument, we will borrow Boder and Gardiol's (1993) words in his definition of knowledge as follows: "any information integrated into its use, that is, comprising the elements relating to its use and to its processing becomes knowledge" [our translation]. Two cognitive sub-tasks can be elicited from this definition: mastery of the information – corresponding to the exploration, collection and restructuring stage – and the integration of the information in a given use – equivalent to the stage of presenting practical use of information in a project, and on the feedback on the activity. This last stage specifically takes account of learners needs through the significant use of information in a project that they themselves have chosen to design and carry out.

We therefore propose a collaborative learning process which is centered on two cognitive sub-tasks (the collection, exploration, and restructuring of information; presentation of results of the work and a feedback on the activity), and which also guarantees explicit management of these requirements.

4.2 The Implementation of the Process Using Computer Conferencing

Our pedagogic use of computer conferencing is guided by a philosophy of mutual assistance and positive interdependence adapted to collaboration; but it also takes into account constraints that attend to the technology as well as peculiarities of distance education for adult learners. Our experience about the implementation of the collaborative process is based on long practice and profound knowledge of what "works" and of what "does not work" when applied to our distant learners.

The peculiarities of distance education for adult learners do not always make it possible to apply the factors that ensure success of collaborative work. Such peculiarities include group competence, the sharing of responsibilities and concentration on a common goal. Without questioning the importance of collaboration among distance education learners, it should also be noted that this reality calls for a flexible functioning of computer conferences requiring participation, cohesion, a sense of belonging and the desire that all should complete their activity, rather than having interdependence as the pivot of functioning.

We know the factors that guarantee the success of computer conferences as regards technology. They have to do with the goal and reason for holding conferences as well as the number of conferences organized, the coordination of discussion and the size of the groups. Conferences must have a single goal, clearly defined and understood by all. Their number must not be increased at the risk of provoking confusion among participants and reducing exchange. Consequently, we must make a judicious breakdown of conferences and take into account the achievement of specific sub-tasks within each distinct conference.

Thus, to implement the collaborative process, there should be an initial conference devoted to the exploration, collection and structuring of information; the second, to the carrying out of the project, and the third, to facilitating the overall management of activity.

In most conferences, a moderator, who should be a competent and an alert specialist, is required in order to coordinate and lead discussions and achieve optimum group productivity. But then, as concerns the moderation of discussions, we learn from works on collaborative learning that authority is shared within the group in a spirit of equality. This gives rise to the emergence of a paradox between the requirements of the communication through computer conferencing and those of collaborative learning. Measures should be taken to correct this paradox. In conferences where the participation of a moderator is required, the moderator must coordinate discussions while, at the same time, encouraging participation and providing motivation. He must avoid being identified with authority; rather, he should systematically encourage exchange among participants. In conferences where a moderator is not needed, and in order to enhance learner responsibility, commitment towards the group, communication skills as well as social and inter-personal skills, learners will be invited to take turns in coordinating conferences devoted to the carrying out of their projects.

Lastly, the number of participants may vary, but it must be admitted that at least fifteen persons will be needed, in order to ensure that there is a minimum of messages in a discussion conference. Contrary to what is recommended for face-to-face collaborative learning, here, it is not appropriate to form exclusively small groups for conferences. The number of participants must vary with the task to be accomplished during a conference. Conferences that aim at exploring, collecting and structuring information are very suited for exchange in big groups, whereas as regards the carrying out of a project jointly, small groups would be required in order to facilitate coordination.

To make the process operational, we have distinguished at least three types of teleconference, and established a relationship between them and the three sub-tasks of collaborative learning. The sub-task on exploration, collecting and structuring of information is carried out almost exclusively within the context of discussions, whereas the sub-task relating to the carrying out of the project is, to a greater extent, the fruit of work that is jointly done. The sub-task on management generates exchanges to set the physical, temporal, spatial and organizational parameters of the process. Hence, we would have "telediscussion-type" conferences (as we have already pointed out), "telework-type" teleconferences and "telemanage-ment-type" conferences as well. These three conference types are characterized in term of the nature of the task or goal, the size of the group and coordination.

The objective, which is shared by participants of a telediscussion, is to set up a project, whether joint or not, that will be carried out outside the conference; telework, for its part, aims specifically at carrying out a joint project within the framework of the conference. Telemanagement involves the communication of instructions to the group, the adjustment of the organizational set up, the making of logistical-type decisions and the follow up of the entire collaboration process. As regards the size of the groups, a large number of participants would be required for a telediscussion so as to ensure more enriching and productive exchanges. But for telework, much smaller groups would be required in order to facilitate the carrying out of the project. Telemanagement is ideally conducted with a mid size group, composing several small work groups; this approach is preferred as it creates some cohesion among learners in order to preserve the sense of belonging to a larger community.

Since a larger number of participants is preferable for telediscussion, the presence of a moderator is required. The dedication of telemanagement to the follow up of the process as a whole, benefits from its being coordinated by an expert moderator. The telework groups can, because of their small size, ensure their own coordination of discussions. Table 1 contains the characteristics that, from our point of view, distinguish telediscussion, telemanagement and telework.

Table 1. Types of teleconference and their characteristics

Types of teleconference Characteristics	Telediscussion type	Telework type	Telemanagement type
Tack	Set up a project, whether joint or not, to be carried out outside this teleconference	Carry out a joint project within the teleconference	Communicate instructions, decide on logistics, follow up of the entire process
Size of group	Larger group (20 to 30 at most)	Small group (3 to 5)	Mid-size group (20 at most)
Coordination	By expert moderator	Self-coordination	By expert moderator

4.3 Summary

The main operational variables of the distance collaborative learning process through conferencing can be summarized as follows.

The Virtual Environment

Here, there are three conferences which correspond to the three principal sub-tasks of the collaborative process (Table 2). These conferences, which are interdependent and complementary, are all involved in the entire process. Their reduced number is intended to avoid the confusion and dissipation of messages. Among them, two operate in the form of a discussion and the other, as a common effort to carry out a project.

Table 2. Sub-tasks of collaborative process and types of teleconference

Sub-task related to the process	Type of teleconference
Exploration and structuring Management Carrying out of joint work	Telediscussion Telemanagement Telework

Size of Groups

In conferences that bring together a large number of participants, the loss of a feeling of belonging and of commitment towards the group is felt. However, the content often turns out to be very rich because of the variety of contributions. Moreover, conferences in small groups, though easy to manage, hardly ever have the minimum required messages, except in the case of telework. The solution to this problem is to adapt the size of the group to the type of conference. Those that carry out the exploration and structuring of information can take larger groups (20 to 30 participants at most); those that focus on the management of activities cannot take groups that are too large (30 at most); and those in which the task is done jointly (work, an assignment, a project) must necessarily be reserved to small groups of 3 to 5 participants. It is within these small groups that the learners have to coordinate work by themselves.

The Composition of Groups

According to our practice, the formation of large groups is randomly done, but the formation of small groups can be negotiated. Heterogeneity is not needed in all cases. This depends on the objectives and theme of the conference.

The Task or the Objective

In telediscussion, the task or set objective is the contribution of experience and information, the clarification of ideas put forward, some deeper understanding and the putting together of ideas and points of view within a common perspective. Discussions are conducted with the aim of enabling each participant to choose pertinent and significant information so as to structure it according to the purpose of discussion and integrate it into his or her own process. In the telework conferences, the exchanges are geared towards the joint performance of a task through teleworking. Finally, telemanagement discussions are aimed at dealing with the planning, organization logistics of the collaborative process.

Because of the very different nature of these three types of conference, the messages are not expected to have the same content or to develop in the same manner. The interactions are completely different in both cases. One should not therefore expect the same quality and level of interaction in all the conferences or make generalizations based on these criteria. This explains why, after analyzing transcripts of adult distance education learners, Mason (1991) and Henri (1992b) came up with different results. The former found little interaction, while the later found a lot more. The difference can be explained by the fact that these analyses do not take into consideration the specific nature of the task, the structure of the learning process, the particular virtual environment and the types of learners. The latter differ in their cognitive maturity, their familiarity with the medium, the organizational culture, the school model used as a reference and, this irrevocably influences their participation in the conference.

The Moderator's Role

The moderator plays the role of facilitator during a telediscussion. He is expected to understand what is being said in relation to the theme discussed, to seek and identify the social and cognitive processes used by learners, to make a diagnosis of problems with the content and the process, to plan and implement appropriate support interventions. In all, the main role of the moderator is to offer both individual and group support in the learning process. The coordination of telework can easily be handled by learners themselves. Instructions and coordination guidelines should therefore be sent out and models of "good" coordination should be provided by expert moderators. Telemanagement requires specific know-how, that learners hardly have. The moderator, in this type of conference, gives instructions and advises on the methodology. He also forms the groups, ensures the follow up on their work, revises and adjusts the unfolding of the collaborative process based on the dynamics emerging from each group.

Remarks

It can not be claimed that the tables and observations presented herein cover all types of conference, particularly those involving, for instance, decision making. They are simply models representing studies we conducted in the course of our research. Nevertheless, they have helped us develop a more generic model of conferencing. Further studies should lead to the development of some sort of taxonomy of teleconferencing that will help to classify and rationalize a series of groupware (tools) such as electronic mail, bulletin board, group scheduler, group decision-making support system, collaborative authoring tool, etc.

5   Developing Tools and Support for Optimizing the Collaborative Learning Process

The amount of information currently available on the collaborative learning process in conferencing does not allow distance educators and trainers to maximize their use of the transmitted messages. In addition, there is a lack of appropriate means with which we can process and analyze the abundant information contained in these messages. There is therefore an urgent need to come up with a method of decoding participants' exchanged messages to bring out the different meanings which contribute to learning. These studies should be carried out with the aim of providing trainers with user-friendly tools which allow them to locate meaningful learning items in the knowledge contents and processes. We should also elaborate rules so that new understanding acquired through the analysis of messages could be used to identify and design learning support roles for individuals and groups.

In our research, we have especially examined three instances of the three types of teleconference. First, a telework type in which learners, in small groups of three, work together in self-coordination to write a short opinion text. Secondly, a telediscussion type, where learners, with the help of a moderator, discuss what they watched on video in a thirty-person teleconference discussion. They then individually prepare, later or during the same period, to take an examination which aims at validating what was learned using these different media. Thirdly, a telemanagement type, in which small groups formed from among the twenty learners participating, develop a work schedule and make a progress report of the whole group.

To begin building an approach to message analysis, we have chosen to dwell on the telediscussion type of conference, as it appears to us that telework and telemanagement are better understood in current studies and now have a number of enhancement tools, some of which are even available commercially. However, a great deal of what we are about to say covers these three cases.

Our approach includes the development of a model and some tools of analysis that essentially aim at:

– reducing the difficulty of obtaining qualitative data on collaborative process that researchers in this area are faced with;
– offsetting the absence of evaluation tools intended for participants in teleconferences;
– providing learners with the opportunity to benefit, at all times, from collaborative learning, through the sharing of knowledge with one another.

To carry out the analysis of messages presumably requires a considerable amount of data essential to this kind of work. The researcher must turn to the fields of linguistics, psychology, sociology, management, cognitive engineering and computer sciences, while at the same time ensuring the coordination of these different contributions and their convergence towards the pursued goal, which is both of a pedagogical and collaborative nature.

5.1 Analyzing Messages

Our aim is to develop a general model of analysis adaptable to the various types of conference. This model can be stated as follows:

– a tool for describing contextual elements of each teleconference;
– an author data base describing participants in one or more teleconferences;
– a tool for describing each conference including:
  – quantitative data dealing with the performance of participants in a particular conference;
  – aspects of the "how" of what is said (language, cognitive functions, processes and strategies, interaction, etc.) and the strategies used to say it;
  – the "what" of the same speech, that is, the transmitted knowledge.

In our model, we integrate the analysis of the message content, and the pattern developed from a conference using data about teleconference contextual elements, as well as data from the authors who participated in it.

The Description of Contextual Elements of a Teleconference

This description is of interest mostly to researchers and designers, as it contributes to the understanding of remarks made about a given set of messages or about a teleconference as a whole. Its main importance can be seen from a perspective of comparison between the analyses of many similar conferences, or between different conference types which are all directed at the same group. We consider mainly the following:

- the social structure in which the conference takes place (group portrait, organizational context, etc.);
- the technological environment available to the participants (conferencing software, mainframe, communication software, hardware, technical support, the temporality of exchanges, etc.);
- the task to be completed and the objectives to be attained.

The Author Data Base

The data base, for its part, considers the sender of messages as a teleconference participant associated with a role: coordinator or ordinary participant. In our training telediscussion context, the ordinary participant is a learner whose social characteristics include age, place of residence, profession, an academic record, progress within a study programme, level of knowledge of the medium and quantitative data on the number of messages, separately or in relation to other participants. Except for the last one, these are all factors which will not vary for a given learner, throughout a series of conferences. These data can, in conjunction with other data, be of interest to both the researcher and the trainer.

Approaches for Describing the Content of each Conference

Analyzing the content of a given conference can be done in at least three ways:

- what is said about the theme of the discussion;
- how it is said;
- what are the strategies used to say it.

Within the framework of education or training, an initial analysis of messages should make it possible to identify in them the knowledge presented on the object of the discussion (*what is said*). This wealth of knowledge produced by participants during the teleconference, is a source of mutual enrichment for each one of them, and may serve as a collaborative support feature. At another level, the analysis relies more directly on how communication works in teleconferencing (*how it is said*). Three factors are considered: the intrinsic value of the deliveries, their level of interactiveness and their relevance. The extent to which the learner participates in the exchanges, with whom and with how much success, can be appreciated through this kind of analysis. It also makes it possible to find out if new knowledge has been produced, as well as to specify how information has been processed. Thus, the accuracy and coherence of the ideas expressed as well as the relevance of the subject can be argued. Finally, the analysis dwells on the strategies that participants use and those which are exteriorized to process information relating to the theme discussed (*what are the strategies used*). The implementation of processes emanates from strategies which themselves rely on skills that the learner uses and masters fairly well. Aside from social and interpersonal skills, there are generally two types of strategy and skill which enhance the learning process: cognitive strategies and metacognitive strategies (cf. Deschenes et al. 1992)[1]. Based on the medium, on the collaborative approach it encourages, as well as the distance education context, one should add organizational-type strategies and skills which learners develop in this area.

Although the three approaches proposed to analyze content are of interest to researchers and distance educators, only the first (*what is said*), at least in the current state of affairs, is capable of satisfying learners' needs. Likewise, even distance educators and researchers differ with respect to the depth of the analysis required. Designing the tool should take these different needs into consideration.

Moreover, there is no use, in this context, except for researchers, to propose an instrument which will be applied after the fact ("post-mortem"). The analysis and re-use of submitted elements in the conference should thus be made while the latter is going on, in support of the collaborative process.

5.2 Presenting a Method for Analyzing Messages

We carried out message analysis by:

– determining the relevant basic units to be analyzed;
– developing analysis grids which identify, in the text of the message, the knowledge and the processes made use of by the learners; and
– using a software for textual data processing; this software makes it possible to:
   – extract a typology of messages, compare these messages to sample messages to evaluate participants' performance, view them from numerous facets, as the researcher wishes, and
   – set up a knowledge base that is accessible to all, particularly to learners, throughout the conference.

---

[1] In the literature, we also find affective-type strategies which allude to motivation, to the interest and emotional state of the learner with respect to the assigned task and the learning situation in general. We include this type of strategy in the metacognitive dimension, when dealing with the learner himself, and in the organizational dimension when dealing with the group.

Work Units

In terms of analysis, the message is a unit which is both too big to be codified and too small to cover the interactions between participants.

We have therefore brought in new units, partly taken from pragmatic linguistics, from the Geneva School, in particular (Roulet, 1985, Moeschler, 1985, 1989). We therefore consider:

- exchange as the smallest unit of interaction involving more than one participant;[1]
- delivery (or message) as the greatest unit controlled by a speaker in a speech event;[2]
- the speech segment as the smallest unit of delivery, linked to a single theme, directed at the same interlocutor (singular, plural or indefinite), identified by a single type (linguistic), having a single function (in relation to the strategies).[3] The basic unit to be analyzed and codified is the speech segment.

The Analysis Grid

As shown in Fig. 1, speech segments are defined based on the fact that they have characteristics, functions and content. Each of these three defining elements are represented in a series of trees (Figs. 2–9) that are used as the grid to analyze the speech segments.

The characteristics. The characteristics of the speech segment include:

- its coordinates (individual or relational);
- its type (at the pragmatic linguistics level); and
- its relative value with respect to the theme and objectives of the conference (relevance). See Fig. 2.

The relational coordinates of a speech segment, as shown in Fig. 3, are especially important to study, since by identifying and qualifying the links between the statements made by various participants or still by the same participant, we get a portrait of the exchange dynamic (interactivity) and the relative position of everyone in the flux, and as well as the facts that are necessary in the analysis of the coherence of each participant's delivery.

As for the linguistic type associated with the segment (see Fig. 4), it corresponds to the fundamental uses of language (assertive, directive, expressive, etc.) and the functions associated with it give us information about the cognitive attitudes of the user (Frechet, 1992). As an example, "say, react, then believe,

---

[1] An exchange involving only two rounds of speech is a minimal exchange.

[2] A one-way delivery has only one speech segment, the complex delivery has more than one speech segment.

[3] The speech segment is a unit which is bigger than the speech act as it is defined by Austin (1962) and Searle (1969; 1979).

think and perceive" are, on a privileged basis, associated functions of the assertive-type segment.

The relative value of the content expressed is establish with respect to its relevance to the theme of discussion and to the objectives of the conference (see Fig. 5).

The functions. The analysis of functions is directly related, as we saw earlier, to the analysis of skills and strategies, revealed by traces left in the speech. Trees used to conduct this analysis are reproduced in Figs. 6–9. One difficulty is found essentially in the diversity of approaches adopted to study such diverse but puzzling facts as social, cognitive, metacognitive and organizational functions and cross check points of view. The latter two cases especially provide an interesting example of the need to develop approaches based on what has been called coordination science (Malone and Crowston, 1993).

The content. The content of the segments is analyzed to constitute a knowledge data base designed in the form of a map representing an area studied by the participants and which is the basis of their discussions. This is entirely appropriate and even expected, considering that the trainer has indeed some way of verifying, at least within the discussion, what has been picked up from the ideas being circulated, the level of assimilation and the transfer capabilities of the learners. However, this does not prevent novelty, and the integration of a new idea, additional information, a broadening of horizons warranted by exchange, in the knowledge tree. The software we are using makes this possible and we hope to develop strategies with the goal of better exploring the collaborative nature of enterprise. Ideally, the discussions could lead to a prompt reorganization of the initial structure, as the telediscussion is being carried on.

The Software

The analysis is conducted and the data processed with ACTIA[1], a system that uses a language called FX (Plante, 1993). FX language allows for the development of knowledge-based systems, with an especially fine touch. One of the problems with these systems, in fact, is often the difficulty of meeting the autonomy requirements of the knowledge units and that dealing with expressing uncertainty. FX offers an alternative solution by applying an analogical rather than a deductive calculation, and showing the analogy and difference between a collection of messages, or speech segments, using multi-faceted comparisons. A tool like this is especially effective in making a typicality judgment which, in our case, means solving the postulated problem.

---

[1] Prototype produced by C. Ricciardi Rigault, F. Henri, L. Dumas (ATO*CI Center, UQUAM) 1993.

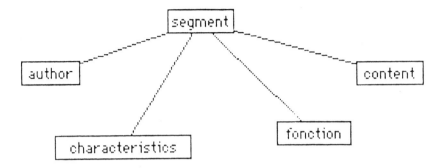

Figure 1. Definition of a speech segment

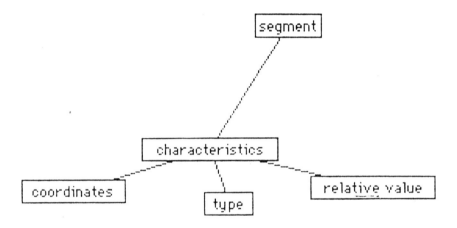

Figure 2. Characteristics of a speech segment

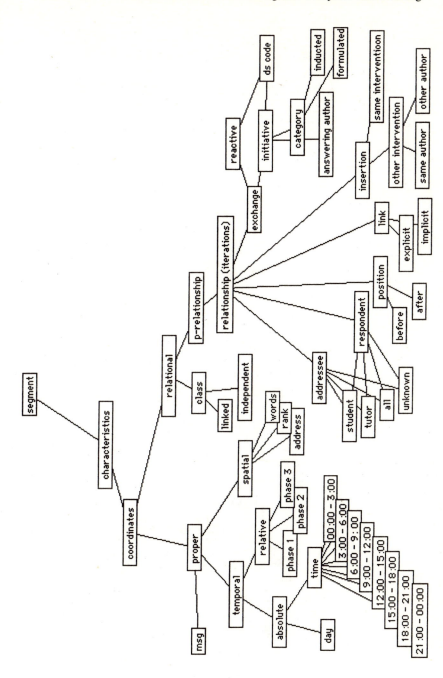

Figure 3. Coordinates of a speech segment

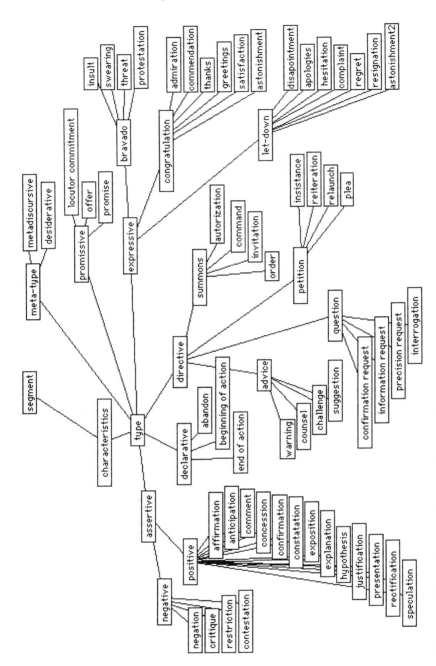

Figure 4. Linguistic type of a speech segment

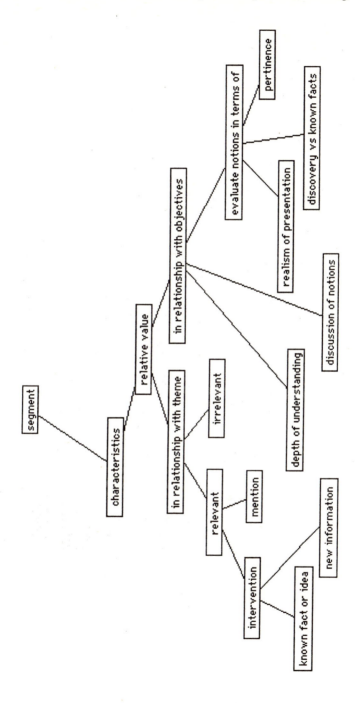

Figure 5. Relative value of a speech segment

68   F. Henri and C. R. Rigault

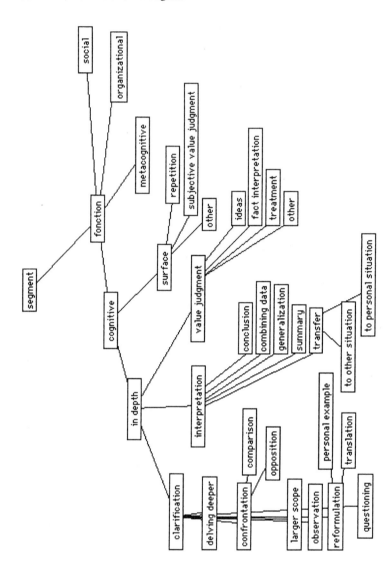

Figure 6. Cognitive function of a speech segment

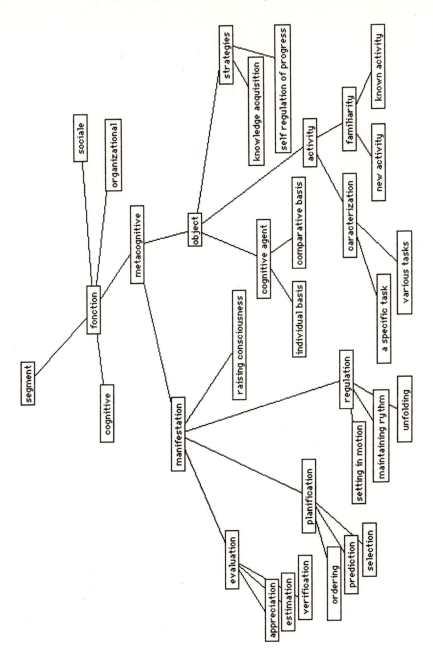

Figure 7. Metacognitive function of a speech segment

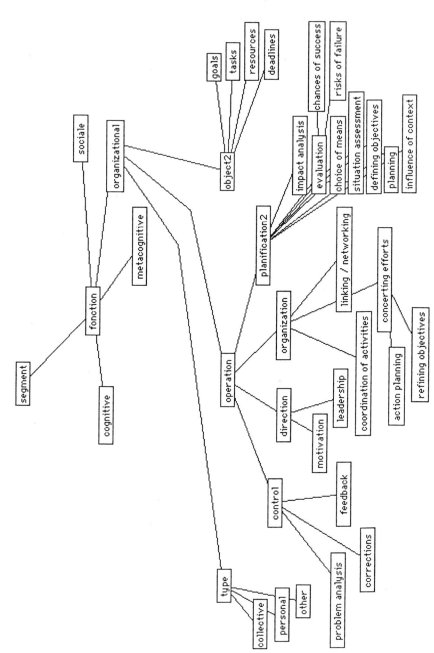

Figure 8. Organizational function of a speech segment

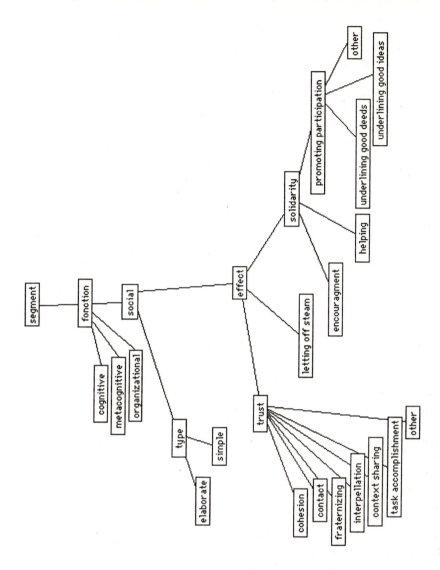

Figure 9. Social function of a speech segment

The virtual FX trees, developed from the analysis grids, enable us to describe the speech segments under the form of real trees which are put together in a base. It is therefore possible to compare a given virtual tree to a given collection of real trees (speech segments) of the base. This will result in establishing the degree of resemblance between the FX tree and the base (real trees describing speech segments). We could, for example, compare a prototype tree with a real tree base and obtain, as a result, a degree of resemblance between the prototype tree and the trees in the base, that we have called the salience coefficient. Salience is calculated on the basis of two principles: first, it is not driven by any external algorithm (the tree drawn by the user is initially responsible for the results obtained); second, base entries are graded proportionally to a typicality criterion (for example, peculiarity or redundancy). It is worth noting that during the setting up of a prototype tree, explaining a coefficient of certainty is not required at all, although making simple typicality judgments is. The degree of association is not assigned a number, but simply, a declaration is made that a set of traits is typically associated, and it is the system that is responsible for working out the production of this typicality in the base. This way, we can identify the trees in the base in a totally independent manner.

5.3 ACTIA Allows for the Appraisal of the Process

A friendly interface invites the users of ACTIA – researcher, learner, educator or trainer – to encode the content elements relating to various aspects of the analysis. Coded data processing produces results and presents them in a manner that anyone can easily decipher. First of all, it helps the moderator to evaluate learners' participation in the conference and their degree of collaboration. The strategy we have used for this purpose is to develop a prototypical representation (virtual tree) of what we could describe as "good collaborative speech segment" when a conference is conducted within a collaborative framework. ACTIA compares this prototype tree to an all-participant speech segment representation, thereby helping to determine the degree of proximity between actual speech segments and the ideal representation. The system otherwise indicates at what point concrete and individual speech segments deviate from the model. It also helps to make a comparison between a learner's messages and that of his peers, or still, to discover through successive comparisons of one tree with all the others, some kind of message typology. All of this, for example, enables the moderator to react to the organization and the conduct of a given conference.

6   Conclusion

The ideas and study that we have just discussed are a follow-up of work on the analysis of messages begun a few years ago and aimed at achieving greater understanding of the learning process through computer conferencing. Our work was initially motivated by the desire to provide first and foremost researchers, and secondly teleconference moderators with valid and reliable tools that make it possible for strategies manifested by learners to be followed up. (Henri, 1992c). In spite of the limits of the instruments and of method used in our initial studies, results obtained show that it was very possible to bring to light the cognitive and metacognitive strategies used by the learners. Then, we pursued our work, partly to improve on our methodology and to find computerized means of analysis and processing of messages, and partly to reinforce the scientificalness of our approach by refining the description of what we wanted to analyze. Hence, the need to describe, as we have done, collaborative learning, to single out its characteristics and to make its process implementable within the framework of distance education for adult learners through teleconferencing.

Although the initial objective of our work was to develop tools for researchers and distance educators/moderators, we discovered that the analysis of the knowledge content within the message could perhaps generate other types of results useful to the learners. We, therefore, added to the analysis of strategies, the analysis of content with a view to creating a knowledge base that is accessible to the learners throughout the teleconference. This tool would help them to improve on the quality of their questions, to enrich the conduct of conferences and, could facilitate more collaborative learning.

The general model that we are proposing comprises several elements: a tool for the description of context components of teleconferences, an author-data base, and a description tool for each of the conferences. The description tool for each conference is based on analysis grids, an encoding method, a processing software for textual data, as well as means of retrieving stored information (process components and knowledge content). We are pursuing our work to polish our method conceptually in order to make the tools of analysis more dependable, as well as to develop rules of interpretation that are easily applicable.

References

(NATO ASI Series F volumes are indicated by F and volume number)

Abrami P. et al. (1993) Using cooperative learning. Montreal, Center for the Study of Classroom Processes, Concordia University

Austin, J.L. (1962) How to do things with words. Oxford University Press

Boder, A., Gardiol, C. (1993)   Le juste-a-temps de la connaissance dans les organisations. Les Actes du 4ieme Colloque international en informatique cognitive des organisations. Tele-university, Montreal, 4-7 mai

Boomer, G.(1990) Empowering the learner. In: M. Brubacher, R. Payne, K. Rickett (eds.) Perspectives on small group learning. Oakville, Ontario: Rubicon Press

Cooper J. et al. (1990) Cooperative learning and college instruction: Effective use of student learning teams. Dominguez Hills, CA: California State University

Deschenes, A.J. et al. (1992) Les activites d'apprentissage pour des cours concus pour l'enseignement a distance. La Revue de l'enseignement a distance VII(1)

Evans, T., Nation, D. (eds.) (1989) Critical reflexions on distance education. London, New York: Palmer Press

Frechet, A.-L. (1992) Analyse linguistique d'un corpus de dialogue homme-machine, Paris III, thèse

Haile, P.J., Richards, A.J. (1984) Supporting the distance learner with computer teleconferencing. Paper presented at the 15th Annual Convocation of the Northeastern Educational Research Association, Ellenville, NY

Harasim, L. (ed.) (1990) Online education. Perspectives on a new environment. New York: Prager

Harasim, L., Wolfe, R. (1988) Research analysis and evaluation of computer conferencing and networking in education. OISE, Ontario Ministry of Education, Toronto

Henri, F. (1992a) Processus d'apprentissage distance et teleconference assistence par ordinateur: un essai d'analyse. Canadian Journal of Educational Communication 21(1)

Henri, F. (1992b) Formation distance et teleconference assistance par ordinateur: interactivite, quasi-interactivite ou monologue. La Revue de l'enseignement à distance VII(1)

Henri, F. (1992c) Computer conferencing and content analysis. In F90

Henri, F. (1993) Formation à distance, materiel pedagogique et theorie de l'education: la coherence du changement. La Revue de l'enseignement a distance VII(1)

Henri, F., Kaye, A. (1985) Enseignement à distance – Apprentissage autonome? In: Henri, F., Kaye, A. (eds.) Le Savoir a domicile. Pedagogie et problematique de la formation a distance. Quebec. Les Presses de l'Université du Quebec

Henri, F., Lamy, T. (1989) La formation a distance: des choix technologiques et des valeurs. In: Sweet, R. (ed.) Post-secondary distance education in Canada. Athabasca: Athabasca University

Henri, F., Lescop, J.Y. (1987) La communication assistee par ordinateur en formation a distance; vers une strategie d'implantation d'innovation. Notes de recherche. Quebec: Tele-university

Hiltz, S.R. (1985a) The virtual classroom: Initial explorations of computer-mediated communication systems as an interactive learning space. New Jersey Institute of Technology

Hiltz, S.R. (1985b) Online communities: A case study of the office of the future. New Jersey, Ablex

Hiltz, S.R. (1990) Evaluating the virtual classroom. In: Harasim, L. (ed.) Online education. Perspectives on a new environment. New York: Prager

Hiltz, S.R., Turoff, M. (1982) The network nation. Human communication via computer, 4th ed. Reading, MA: Addison-Wesley

Hooper, S. (1992) Cooperative learning and computer-based instruction. Educational Technology Research and Development 40(3)

Hotte, R. (1990) Profil de l'étudiant adulte. Notes de recherche. Quebec: Tele-university

Jacquinot, G. (1990) Les enjeux de l'apprentissage dans l'enseignement à distance. Conference d'ouverture. Congres de l'Association Canadienne de l'education à distance, Tele-university, Quebec

Jacquinot, J. (1993) Apprivoiser la distance et supprimer l'absence? ou les defis de la formation à distance. Revue Francaise de Pedagogie 102, janvier-fevrier-mars

Jarvis, P. (1986) Models of distance education. International Council for Distance Education Bulletin 11

Johansen, R., Vallee, J., Spangler, K. (1979) Electronic meetings: Technical alternatives and social choices. Reading, MA: Addison-Wesley

Johnson, D.W., Johnson R.T., Smith K.A. (1991) Cooperative learning: Increasing college faculty instructional productivity. ASHE-ERIC Higher Education Report No. 4. Washington, D.C.: The George Washington University, School of Education and Human Development

Kaye, A. (1987) Introducing computer-mediated communication in distance education system. Canadian Journal of Educational Communication 16(2)

Kaye, A. (1992) Learning together apart. In F90

Kerr, E.B., Hiltz, S.R. (1982) Computer-mediated communication systems. New York: Academic Press

Laurent, J. (1987) Les étudiants de la Tele-university et leurs projets d'études Enquete sur les "abandons" des inscrits de l'automne 1987. Quebec: Tele-university

Malone, T.W., Crowston, K. (1993) The interdisciplinary study of coordination. Report, Center for Coordination Science, MIT

Mason, R. (1991) Analysing computer conferencing interactions. International Journal of Computers in Adult Education and Training 2(3)

Mason, R. (1992). Evaluation methodologies for computer conferencing applications. In F90

Mason, R., Kaye, A. (1989) Mindweave. Oxford, Pergamon Press

Meunier, C., Henri, F. (1987) Recherche en telematique et formation à distance. Les Actes du Premier Congres des Sciences de l'Education de langue francaise du Canada, Quebec

McCreary, E. (1989) Eliciting more rigorous cognitive outcomes through analysis of computer-mediated discussion. Paper prepared for Improving University Teaching, 15th International Conference, Vancouver

McCreary, E., Van Duren, J. (1987). Educational applications of computer conferencing. Canadian Journal of Educational Communication 16(2), 135-166

Moeschler, J. (1985) Argumentation et conversation. Paris: Hatier

Moeschler, J. (1989) Modelisation du dialogue. Paris: Hermes

Paquette, G., Bergeron, G., Bourdeau, J. (1992) La classe virtuelle: un environnement technologique de formation. Etude realisee dans le cadre du projet "Impact de l'informatique cognitive dans l'enseignement universitaire". Laboratoire d'informatique cognitive et environnements de formation (LICEF), Tele-university, Montreal

Plante, P. (1993) FX. La programmation en faisceaux. RDLC, Centre ATO*CI, Université du Quebec à Montreal

Pradl, G.M. (1991) Collaborative learning and mature dependency. In: M. Brubacher, R. Payne, K. Rickett (eds.) Perspectives on small group learning. Oakville, Ontario: Rubicon Press

Reid, J., Forrestal, P., Cook, J. (1984) Small group work in the classroom. Perth: University of Western Australia

Ricciardi Rigault, C., Henri, F. (1989) Support à l'apprentissage. Actes du Colloque Le transfert des connaissances en sciences et techniques, Université de Montpellier II

Searle, J.R. (1969) Speech acts. Cambridge, UK: Cambridge University Press

Searle, J.R. (1979) Expression and meaning. Cambridge, UK: Cambridge University Press

Roulet, E. (1985) L'articulation du discours en francais contemporain. Bern: P. Lang

Shapiro, H., Moller, M., Nielson, N.C., Nipper, S. (1987) Third generation distance education and computer conferencing in Denmark. Paper presented at the 2nd symposium on Computer Conferencing, University of Guelph, Guelph

4

Interaction and Collaboration in Distance Learning Through Computer Mediated Technologies

M. Felisa Verdejo

Departamento de Ingeniería Eléctrica, Electrónica y de Control, E.T.S. Ingenieros
Industriales, U.N.E.D, Ciudad Universitaria s/n, 28040 Madrid, Spain
E-mail:felisa@horacio.dieec.uned.es

Abstract: CMC technology addresses two key issues in distance education: information transfer and interaction. In this paper we will review some of the potentialities and problems arisen by the use of this technology, as well as how it affects the current distance education framework both from the educational and organizational point of view. The new potential of CMC to support collaborative learning at a distance is illustrated through examples. Metaphors and models for group interaction and coordination are reviewed and main requirements for the supporting software are pointed out.

Keywords: CMC, distance learning, computer-supported collaborative learning, collaborative learning design, collaborative learning evaluation, computer-supported group coordination

1 Introduction

Computer Mediated Communication (CMC) brings together computers and telecommunications, and may include electronic mail, computer conferencing, computer bulletin boards, facsimile, teletex, video text, voice messaging, desktop videoconferencing, etc.

CMC technology addresses two key issues in distance education: information transfer and interaction. In this paper we will review some of the potentialities and problems arisen by the use of this technology, as well as how it affects the distance education framework both from the educational and the organizational point of view.

In Section 2 we will discuss the use of CMC on the current context of distance learning institutions focusing at the university level. Then Section 3 introduces the new potential of CMC to support collaborative learning at a distance and discusses, using examples, principles to design and carry out learning experiences based on collaboration. Metaphors and models for computer-supported group interaction and coordination are further reviewed in Section 4. Finally the conclusion stresses main requirements for the supporting software.

2   The Potential of CMC in the Current Organizational Context of Distance Education

The current organizational context of distance education is centered on stand-alone study using instructional material designed and supported by the staff of a central organization. Tutors are available in distributed study centers and/or in the main institution, to provide help and personal advice. Communication takes place mainly by phone, mail or visiting the tutor.

Many experiments (Mason & Kaye 1989, Paulsen 1993) have been carried out to introduce CMC in this framework and there is general agreement on the potential of these technologies to enhance the support given to students by improving information transfer and communication.

CMC offers new facilities for distance education but also raises new problems, mainly derived from the lack of models, both at the didactical and organizational level, to guide the design, implementation and evaluation of a CMC-supported learning process. There have been various attempts in this direction, for instance (Paulsen et al. 1993) define a framework where dimensions are given in order to characterize the setting of a CMC system for learning purposes. On one hand they consider the universe of CMC sources: people, information and applications, on the other hand, the environment, further defined in terms of constraints, social and staff demands and choices. Among the constrains they consider type of institution, geographical issues, equipment, course time frame, financial support, and institutional resources. For choices, a further division is given between those features characterizing the target group, such as group size, educational level, CMC expertise, computer attitude, motivation and financial status and those specifying the program objectives: subject-matter, formal curriculum or not, etc. Taking into account these features, a range of pedagogic techniques and styles are recommended. Pedagogical techniques, defined as a manner of accomplishing teaching objectives, are described and organized in terms of a supporting communication paradigm, for instance, information retrieval is under one-alone class, while computer conferencing belongs to a many-to-many one.

A similar but more technologically centered approach is proposed by Zorckoczy (1993) with the definition of a *scenario* specified in terms of the following dimensions: Information exchange needs (text, audio, graphics, video ...), user grouping (individual, small group, ...), synchronicity requirements, communication forms (ranging from one-alone to many-to-many) geographically coverage and functionalities (messaging, file exchange, sharing applications ...). From this scenario, guidelines are given in order to select a suitable technology for a distance training application.

In the context of European Distance Learning Universities, several experiences involving the use of CMC have been carried out. The role of CMC differs in these experiments: some of them just incorporate communication facilities in a previously existent course, while others re-designed the course from scratch. As Ellerman et al. (1993) point out, it is not enough to offer the facilities (a computer and a network added to a course) for a sensible use of CMC in distance education. Therefore a first recommendation for distance learning course designers can be stated as follows:

to provide the learner with *an integrated study environment*,
where the technology serves well-described didactical functions.

Integration here refers to different levels: on one hand the communication and computer technology itself, on the other, the relationship between the learning activities and the supporting technology.

For the technological level two aspects should be integrated: the use of multimedia communicative channels, and the tools supporting the student learning tasks.

For the relationship level, the integration should emerge from the didactic of the course. Thus, the student should focus on her learning goals, reachable through various activities, some of them involving traditional instructional material, others can include the use of computer tools, while any of them may require communication.

To go from the experiences to a regular use, a main problem is to scale-up from a few number of students to actual large registration figures. Apart from the technological infrastructure, this implies more human resources and an important change of the organizational model. Distance education has been based upon pre-produced course packages to support individual learning. The social perspective has been peripherical, the main component of this model relies on autonomous learning supported by textual materials with a very limited communication. CMC provides facilities for communication but they can be effectively used only if at the other side there is staff available to respond to the demand.

It seems that in the near future, due to these costs and structural factors, the main activity of a distance learning university student will still remain linked to handling textual course materials. Nevertheless some benefits can be expected at short term in the current organizational context.

CMC allows the automatization of various current tutor's tasks such as the delivery of information, providing questions and answers for standardized consulting, processing forms and students questionnaires, planning and monitoring students tasks, etc. In this way faculty capabilities could be dedicated to provide better study support. But most important, CMC opens a way to collaborate among tutors and students themselves. A network of tutors might supply a richer education environment for students. A group of students working together is a practical way to teach and acquire collaborative skills at a distance.

3   Collaborative Learning

Learning involves active construction and use of knowledge. It occurs when students have to explain, develop or justify their ideas to others. Thus, students might learn from each other when they participate in process-oriented collaboration, solving problems or discussing how to carry out a task. It can be said that learning occurs as a side effect of private and group work collaboration processes, by combining communication and activity among peers.

Collaborative learning is a current trend emerging from a variety of theories, emphasizing the perspective of a social construction of knowledge.

Learning is seen as a dynamic process of knowledge reconstruction and interpretation, carried out through argumentation, debate, and discussion within a learning group.

From this perspective the setting of a collaborative learning environment requires a shift (Clancey 1992, Jonassen et al. 1993), from the media-centered paradigm, linked only to individuals and based on the idea of knowledge as a thing stored and transmitted, to a conversation or dialogue paradigm (Winograd et al. 86), providing support to communicate ideas and coordinate actions.

Taking this point of view, *individual learning as a result of a group process* (Kaye, A., this volume), does not mean excluding other approaches. Learning is a complex phenomenon, and other theoretical frameworks account for the importance of modes such as discovery learning, coaching, or one-to-one tutoring (Vivet, M., this volume). The challenge for an educational designer is to decide the most appropriate combination of methods for a particular learning situation.

A number of requirements for successful collaborative learning have been reported, some derived from collaboration in general: group size, shared goal, mutual respect, individual competency, balanced contribution. Others are specific to learning situations such as type of subject-matter involved or kind of task. For instance (McGrath 1984) describes experiments carried out to compare group achievement through various tasks requiring different skills. He found that for tasks engaging creativity or finely grain-size chain of logic, group performance was worse than individual performance, while in many problem-solving or decision-making situations members of a group performed better than individuals. Benefits associated to a constructivist approach to learning in groups include cognitive gains, social skill development and a significant motivational value. Among the drawbacks are costs concerning the resources invested in group coordination, or undesirable effects such as lack of initiative, conformity, or conflict.

With the availability of CMC technology providing a basis for group communication, the strengths of collaborative learning are appealing for distance education environments and various research experiences have been recently carried out. Next we will describe some of them pointing out two main methodological aspects: principled design and evaluation.

3.1 Design of a Collaborative Learning Experience

As should be the case for any learning oriented environment, the design process has to follow an educational driven approach, thus it is important to stress that applications should not be found by trying to use a particular support technology but by looking for opportunities to work and learn in cooperation and collaboration.

A principled design should first take into account the educational goals, then the activities and pedagogical strategies to attain them, and from there determine the technology needed. For distance collaborative learning, the CMC environment should serve two main functions:

• Mediate communication
• Provide structure for activities in which learning can occur.

To illustrate these ideas in a concrete way we will refer to a case study. Alexander et al. (1993) present an eight-week regular course on *Renewable energy technologies,* the course design started stating the main learning goals. At the end of the course, students should know about the following topics:

• What are renewable energy technologies?
• How can renewable energy technologies be integrated into a nation's energy supply system?
• What are the non-technical considerations affecting the deployment of renewable energy technologies?

In order to carry out these learning goals, three activities are proposed. For the first one, students have to work together to create a document describing renewable energy technology in Europe. In the second one an interactive simulation model is proposed: *integrated energy in Ecotopia*, in which students have to establish their hypothesis and explore the consequences. Finally in the third one, a role playing game is provided, and through it, students have to plan and discuss the setting up of a wind farm.

The activities chosen entail different pedagogic strategies to support students' learning tasks, one of them more oriented to declarative knowledge construction while the two others are more suited for reasoning skills, either procedural knowledge application or argumentation on hypothetical decisions and their effects.

Learning tasks can be characterized by their underlying question. The link between questions and suggested activities for the example above are:

- What are …? joint document creation
- How can …? use of a simulation model
- What if …? role playing in a case study

Students combine private and groupwork along the course's duration. Personal work includes finding relevant information, performing experiments with the model or writing proposals, while the group mode concerns discussing results, commenting other's proposals or responding to requests.

The interactive learning support system integrates a multimedia library with papers, computer models and audio-visual materials, a study room with tools for creating documents, and a meeting room where discussions and activities coordination are carried out. Activities are structured and have to be performed in parallel. For each one, stages are defined, so that group members have to produce outcomes at agreed-upon dates. The system specifically supports collaboration, providing mechanisms for joint authoring, simple voting, checking for mutual understanding or the completion of activities stages.

Many other experiments to explore distance collaborative learning have been reported (e.g. Derycke et al. 1993, Goodman 1992, Hansen et al. 1991, Hsu et al. 1991, McConnell 1992, Rueda 1992); they include other techniques such as debates, joint presentations, brainstorming, group decision-making, project group, peer review, or cooperative problem-solving.

3.2 Evaluation of a Collaborative Learning Experience

A number of recent publications deal with evaluation methodologies either focusing on CMC applications or collaborative learning. For instance, Mason (1992) reviews how a diversity of standard techniques including survey questionnaires, user interviews, empirical experimentation and case study, have been applied in various evaluation studies. For the last approach two kind of methods are further described: quantitative ones, relying on interpretation of computer-generated statistics, and qualitative ones based on content analysis of messages interchanged by members of a group.

Evaluation studies refer to diverse issues, we will focus on those analyzing the quality of the learning process, looking specifically at the collaborative aspects. The first work described next proposes a methodology to study whether or not students are collaborating when solving a problem together, using a common resource in a co-presence setting. The second one deals with the relationships between channels of communication and collaboration, while the third one is centered on group distance learning through CMC, and proposes an approach to structure and analyze member interactions in order to detect patterns of collaboration.

Singer et al. (1988) present a methodology aiming at characterizing how cognitive changes occur when students collaboratively use an interactive graphical model, in the case-study described, a microworld, to learn about the Newtonian concepts of velocity and acceleration.

They propose to carry out the analysis of the learning process at two separate levels: the shared problem space where students perform their actions and the transaction level where communication occurs. For this transaction level they apply discourse analysis techniques to show how communication promotes or inhibits changes in the joint problem space. In particular they propose four criteria as collaboration indicators:

- sharing control of resources
- communicating ideas from each partner to others
- focusing and maintaining attention on similar aspects of the problem
- sharing successes and failures

Their method proceeds in two steps, first an overall analysis looking at whether or not collaboration occurs, and then a more detailed analysis to characterize exchanges in terms of turn-taking versus acceptance of ideas. Pointing out at remote to co-presence differences for collaborative learning (Smith et al. 1991) report an experiment where they observed and compared students working in collaboration in separate places, using electronic mediated communication with students working with physical-copresence. The experiment was carried out using a microworld system that provides real-time sharing of the same animated virtual word for an arbitrary number of users. Four different settings were observed: (a) students in different rooms connected by audio and video links, and the common virtual world in the computer, (b) same situation but the video link removed, (c) co-presence in the same room but students working in different workstations, and (d) students sharing side-by-side a unique workstation. In their analysis they studied: subject's activities, associated discourse and eye-contact behavior, in terms of two set of factors, the first one to discriminate the type of activity and the second one the nature of the same type of activity, further classified into:

- involving the use of the interface
- the problem-solving task
- social interaction
- nature of the activity
- meta-level (generating hypothesis,discussing problem-solving strategies)
- specific
- recovery (conversational repair, recovery form interface errors)

Their main findings relate to how collaboration is affected by two factors: co-presence versus remote setting on one-hand, audio-video-computer versus only audio-computer communication channels on the other.

For the first factor they claim: " we saw nothing done co-present that could not be done in the remote setting".

From the observed correlation between meta-level and specific tasks and non-verbal activity, they conclude for the second factor that "the addition of a video channel to a remote collaborative technology does influence the student's activity by encouraging interactions about the problem". Thus having a visual communication channel encourages interaction in meta and specific activities, and therefore facilitates collaboration. Furthermore, they looked at the relationship between task division, a fact considered relevant for effective collaboration, and proximity, and how this notion may change with technology. They suggest that the topology created by their environment (a shared working space plus the audio-video-channel) offers "an enhanced proximity in which it is possible to be simultaneously side-by-side and face-to-face", and this provides a good support for fluid task division.

In search of a methodology and a tool that could help the educator to understand and give support to distance learning students using a Computer Conferencing environment, France Henri (Henri et al., this volume) proposes an analytical method to carry out an in-depth analysis of the content of messages interchanged among a group of students when engaged in a learning task.

Henri (1992) first considers classifying messages, taking into account five dimensions of the learning process: Participative, Social, Interactive, Cognitive, and Metacognitive. For each dimension a subcategorization is given, where each category is defined and indicators for identifying messages belonging to the class are specified. For instance, the cognitive dimension includes elementary clarification, in-depth clarification, inference, judgment, and strategy. The method proposed consists on breaking messages on "units of meaning", each unit pertaining to one of the categories, so that the analysis result is represented as a matrix showing the structure of messages in terms of the dimensions they contribute to. A set of tools allows the educator to carry out this analysis, and provides for further operations, for instance to evaluate learner's participation, or to compare prototypes of collaborative patterns against the current production. The aim is to understand the group process so that the educator could help or facilitate a more collaborative learning.

4   Metaphors and Models

Members of a distance learning group are individuals carrying our their own work using the resources provided, participating in activities and discussions with others, using the communication facilities. For a collaborative approach the group is a central concept, i.e. a well identified collection of learners who are involved in the satisfaction of common learning goals. As Bannon et al. (1991) signal, a group exists when its members perceive themselves as a *we* , in this sense it is of great importance to capture and visualize the group as an explicit entity within a CMC framework. Up to now, the most unifying metaphor proposed, is the virtual learning center, including rooms with different functions: a multimedia library, a study room with various tools such as joint editing, and a meeting room where groups can have discussions, work together and coordinate activities.[1] As mentioned in the examples above, it is important for collaborative work to explicitly represent two separate but interrelated spaces: the shared object space where actions take place and the conversation space where interactions are carried out. A key issue for the first space is to provide a shared focus, while for the second one it is to support coordination.

Significant work on coordination, both on models and support tools, has emerged in the field of Computer-Supported Cooperative Work. Since these models are inspiring the design of collaborative learning environments, we will refer to them briefly.

Coordination aims at managing interdependencies between activities performed by actors to achieve a goal. The questions addressed (Malone 1990) include: how overall goals can be subdivided into actions, how actions can be assigned to groups or individuals, how resources can be allocated among different actors, and how information can be shared among different actors to achieve the overall goals.

As Rodden et al. (1991) point out, collaboration is a way to solve problems of a very different nature, some of them require a fixed and well-defined group coordination while others need a very flexible and non-regulated interaction. For structured activity, a detailed description of the cooperation can be given, so that it can be modelled and incorporated in a computer program. In this way a system can supervise and facilitate group task completion.

There are a number of models that allow the *explicit* representation of coordination procedures. Three main approaches have been proposed, they present different views for analyzing the problem:

- Activity/task oriented models focus on the way coordination integrates roles and activities. For instance the AMIGO Activity model (Danielsen et al. 1986) specifies the participating roles, the messages they can used to communicate, the operations and the rules stating which operation can be performed on which object by whom.

- Conversation oriented, analyzes group actions based on the language/action theory (Winograd et al. 1987). It makes explicit a common structure for conversation within the group. The key distinctions in this structure are linguistic actions: requests, promises, assertions and declarations.

- Semi-formal structuring, dealing with knowledge represented in such a way as to combine formal processing with informal one. For instance (Lai et al. 1988) propose *objects* to represent "passive information" such as messages and rule-based *agents* for processing information automatically on behalf of their users.

---

[1] See (Derycke, A., this volume) for the discussion of this metaphor from the user interface point of view.

Systems based on an explicit representation of the coordination process not only provide automatic support for group task completion, but also allow members of a group to tailor the system to their own structure and cooperation rules.

Explicit models have been proposed for distance learning groups either to support tutors (Verdejo et al. 1991, Simone 1993) dealing with a large number of students in tasks that could be standarized, such as advising students on common problems, giving information about courses, checking and monitoring student progress, or to support teacher ' tasks related to others academic duties involving more elaborated cooperation procedures. For instance in (Cuena et al. 1993) the system plays a mediator or facilitator role within the group.

Some collaborative learning environments offer semi-structured communication, for example (Alexander et al. 1993), where special types of messages threads are defined, including conversation, commented-document, and virtual-cycle. The last one can contain agreed-upon deadlines and is used for coordinating group activities.

When the control of the coordination is embedded into the system without any inspectionable representation it is called an *implicit* control mechanism. Conferencing systems usually offer this kind of approach: in asynchronous mode through a moderator (Palme 1992), who has the rights to modify the structure or content of the conference, while for synchronous mode a floor algorithm (see e.g. Derycke & Vieville 1993) specifies who has access to a shared space at any moment.

Finally there are systems, as (Smith et al. 1991), that do not provide any control mechanisms over shared spaces, so that coordination and collaboration has to be mutually agreed and carried out by other means.

More prescriptive systems impose constrains to participants but also can give more support for group task completion in highly structured activities. However, informal and direct peer interaction is a key feature for many collaborative learning situations. To summarize, it seems that a single set of mechanisms is unlikely to be suitable for distance learning environments, and a flexible approach should be beneficial. It is still an open question whether the design of collaboration-supporting systems is better served by a generic approach, or by a variety of contex- and problem-specific tools for particular user group needs.

5   Conclusion

It is important to realize that there is not yet a coherent body of theory in the area of computer-supported collaborative learning, however there are emerging theories and concepts that have been applied, proving the potential of the approach. The development of environments specifically built for distance collaborative learning is still in its infancy, but based on the experiences already carried out, there is a broad consensus among researchers on how to approach the design and construction of the computer-support platform. We will end by pointing out some of the main requirements:

- **Integration**, computer support would have to provide integration at two levels: in the use of communicative channels, and with work and learning activities applications.
- **Adaptability** towards communicative processes, because group processes and collaboration are dynamic and therefore subject to changes
- **Learner-centered approach** to the control issue. The organizational context need to be captured but the many different forms of cooperation need to co-exist.

References

(NATO ASI Series F volumes are indicated by F and volume number)

Alexander,G., Lefrere, P., Matheson, S. (1993) Towards collaborative learning at a distance. In F133
Bannon, I.J., Schmidt, K. (1991) CSCW: Four characters in search of a context. Bowers, J.M., Benford, S.D. (eds). Studies in computer supported cooperative work, pp. 3-16. Elsevier
Clancey, W. (1992) New perspectives on cognition and instructional technology. In F91
Cuena, J., García-Serrano, A., Verdejo, M.F. (1993) The role of knowledge based systems for automatic coordination in distance learning. In F133
Ellerman H., Schellekens A., Willibrord H. (1993) An experimental network-mediated study support system. In F133
Danielsen ,T., Panode-Babatz, U., et al. (1986) The AMIGO project: adavanced group commnunication model for computer-based communication environment. Proc. of CSCW 86, Austin, TX
Derycke, A., Vieville, C. (1993) Realtime multimedia conferencing system and collaborative learning. In F133
Flores, F., Graves, M., Hartfield, B., Winograd, T. (1988) Computer systems and the design of organizational interaction. ACM Transactions on Office Information Systems. April, pp. 153-172
Goodman, F.L. (1992) Instructional gaming through computer conferencing. In: Waggoner, M. (ed.) Empowering networks: Computer conferencing in education. Englewood Cliffs, NJ: Educational Technologies Publications
Hansen, E. et al. (1991) Computer conferencing for collaborative learning in large classes. Division of Development and Social Projects, Indiana University
Henri, F. (1992) Computer conferencing and content Analysis. In F90
Henri, F. Rigault, C. (1996) In this volume
Hsu, E.Y.P., Hiltz, S.R. (1991) Management gaming on a computer mediated conferencing system: A case of collaborative learning through computer conferencing. Proc. of 24th Hawaii International Conf. on Systems Sciences, pp. 367-371

Jonassen, D., Mayes, T., McAlesse, R. (1993) A manifesto for a constructivist approach to uses of technology in higher education. In F105

Kaye, A. (1996) In this volume

Malone, T.W., Grant, K.R., Turbak, F.A., Brobst, S.A., Cohen, M.D (1987) Intelligent information-sharing systems. Comunications of the ACM, 30, 390-402

Lai, K., Malone, T.W., Yu, G. (1988) Object Lens: A spreadsheet for cooperative work. ACM Transactions on Office Information Systems 6(4), 332-353

Mason, R. (1992) Evaluation methodologies for computer conferencing applications. In F90

Mason, R.D., Kaye, A. (eds.) (1989). Mindweave: communication, computers and distance education. Oxford: Pergamon Press

McConnell, D. (1993) Learning in groups: some experiences of online work. In F133

McConnell, D. (1992) Computer mediated communication for management learning. In F90

Palme, J. (1992) Computer conferencing functions and standards. In F90

Paulsen, M. (1993) Some pedagogical techniques for computer-mediated communication. In F133

Paulsen, M., Barros, B., Busch, P., Compostela, B., Quesnel, M. (1993) A pedagogical framework for CMC Programs: a work group report. In F133

Rodden, T., Blair, G. (1991) CSCW and distributed systems: The problem of control. In: Bannon, L. Robinson, M., Schmidt, K. (eds.) Proc. of 2nd European Conference on CSCW, pp. 49-64. Kluwer

Rueda J. (1992): Collaborative learning in a large scale computer conferncing system. In F90

Simone, C. (1993) : Supporting collaborative dialogues in Distance Learning. In F133

Singer, J., Behrend, S.D., Roschelle, J. (1988) Children's collaborative use of a computer microworld. In: Proc. of CSCW 88, pp. 271-281. ACM

Smith, R., O'Shea, T., O'Malley, C., Scanlon, E., Taylor, J. (1991) Preliminary experiments with a distributed, multimedia, problem solving environment. In: J.M. Bowers, S.D. Benford (eds.) Studies in computer supported cooperative work: Theory, practice and design, pp. 31-48. North-Holland

Verdejo, M.F., Abad, M.T. (1991) Human-human communication in an open distance learning environment. In: Proc. of 8th International Conf. on Technology and Education

Vivet, M. (1996) In this volume

Winograd, T., Flores F. (1986) Understanding computers and cognition: A new foundation for design. Norwood, NJ: Ablex

Zorckoczy, P. (1993) Educational scenarios for telecommunication applications. In F133

5

Interface Issues in Advanced Educational Technology Based on Computers

Alain C. Derycke

Laboratoires TRIGONE & LIFL, Institut CUEEP – Université de Lille I
59655 Villeneuve d'Ascq Cédex France
E-mail : Alain.Derycke@univ-lille1.fr

Abstract: Rapid progress of technology and design methodology for Human-Computer Interface will have an important impact on the future of Advanced Learning Environments. In order to set up a first research agenda and to derive the interface issues two viewpoints will be developed: one relative to a reframing of learning with its potential paradigm shift, and one relative to advances in HCI. The research issues for interface of advanced learning environment based on computer will be at the cross-road of these two viewpoints. In order to avoid a unique technology-driven approach it will be take into account contribution of various fields of research about human and technology. A summary is given of the state of art of current HCI research in the perspective of the learning process and of the education context. Special emphasis is put on the design process with its new foundation, its metaphorical issues and its design language.

Keywords: Computer-based-learning, multimedia, multimodale, interface design, cooperative software, multi-users interface

Introduction

The present paper tries to define some of the research issues in the field of Man-Machine Interfaces (MMI), through applications in Advanced Educational Technology. In fact, due to the interest of the author and his background, the focus will be put essentially on the Human-Computer Interface (HCI). The shift from MMI to HCI, in the light of educational applications, is not too restrictive because of the ubiquitous character of the computer now. Even when you interact with a modern photocopier, it is a computer which is in control.

The decision to put the learning process, or rather, the learner, at the centre of the educational system will be justified later.

To set up a first Research Agenda we propose to restrict the problem domain to the Interaction between HCI research field and those of new learning technology based learning environment (see Fig. 1). Even with this simplification it is not an easy task to provide a clear view of the real research issues. The difficulty is not only due to the very broad range of research areas, disciplines, etc., that must be considered but also to the epistemological nature of the Interface, at the boundary of two worlds : the user and the system. It is quite impossible to speak about the interface without speaking about the user, his learning process, his social insertion and about the software, for example, with which he interacts. So our first task will be to define, in our context, the concept of interface that we will consider in a broad meaning.

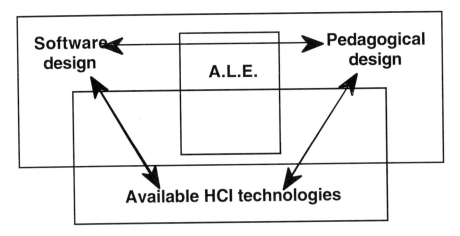

Figure 1. Advanced computer-based learning technology: at the intersection of three research domains

It must be noted that until now there have been few attempts to propose new directions or trends for educational technologies or educational software in regard to the progress of computer telecommunication and educational sciences. Some in-depth views of the potential future are, for example (McLean 1989) for educational software, or (Pea 1992) for distributed multimedia learning environments, or (White 1989) about current trends in education and technology. Our approach is close to those of Pea in the fact that we want to take into account also the results of the research in learning theory, cognitive psychology, and various other social sciences.

1 Interface Concept and Technologies

1.1 Around the Interface Concept

Even if the field of HCI is now well established, there is always some controversy about the meaning of the interface concept. Of course, the interface is always viewed as a transition between the human world of the user and computer world (abstract, virtual, symbolic?). The views diverge about the nature of this transition : from abrupt, interface as a surface, to smooth, interface as a continuum or a layered ("onions") system. For a discussion about the unity of the "interface" concept see (Kuutti & Bannon 1993). For the clarity of the following text we will use Interface, and User Interface (UI) in a broad sense, integrating different viewpoints, from hardware to social and organisational viewpoints, without assuming a clear separation of the "interface" from the underlying "functionality".

Our assumption is that there is an inevitable interwinning between interface and "functionality". The functionality of the system, in our case, is shaped by the learning theories, explicit or latent, espoused by both the designer and the user.

However, this separability is real at the implementation level, due to the line of demarcation between the psychologists, educators and the programmers. This is enforced by the models and tools, such as the User Interface Management System (UIMS) or the Graphical User Interface, which are now popular, and which have spread diffused with the rapid growth of systems such as Windows or X-Windows.

1.2 About Learners, Users and Tools

First of all it is interesting to see that in the field of HCI there is a strong interest in the user as a learner. The "learnability" of the user interface is one of the major concerns of interface designers. Even if the user is a professional interacting with the system in the workplace, this seems to be a major design issue. Brown and Duguid (1992) state that *"Design for the workplace must involve design for learning"*. This reflexive nature of user and learner is of course also present in the field of educational technology but we must consider the learner as a user from the interface viewpoint and as a learner for the system's functionnality and activities viewpoints. However, in most of cases, learning of the interface (how to master it) interacts with the whole learning process.

Secondly, several modern approaches of HCI design and implementation are conducted in a tool perspective: systems and software are designed to provide to the user tools to empower or to augment his human capacities. We will investigate later the potential of such a latent metaphor, tools, for the design of Advanced Learning Environments (Sorensen 1992).

But from the previous human experiences this perspective (design of/and tools) implies three key aspects, as emphasized by Brown and Duguid: Tools are cultural artifacts; Tools provide "affordances"; Tools need to be capable of becoming ready-to-hand.

It is outside the scope of this paper to investigate all the implications of this key aspect, even if the tool perspective is now at the foundation of some very interesting streams of research in educational technology, with, for example the field of cognitive tools for learning (Kommers et al 1992). However the notion of a community of learners and its context (for example the classroom) or the "affordances" of Advanced Learning Environments (ALE) will ground most of our investigations.

The interaction of the users with the tools through or by mean of an interface raises the problem of the usability and of the effectiveness of the system designed. In the educational field the effectiveness (efficiency ?) is in relation to the learning goals and their achievements. Usability is more linked to the quality of the interface. Of course a very usable ALE can be ineffective from the education point of view, due to some drawbacks in its pedagogical design, or to some failure in the context of use. But we are sure that even a well-designed ALE from the learning point of view, will be ineffective if its interface is not usable because of design errors. Again, as in the discussion about the interface concept, there is no clear separation between the usability and the effectiveness of an ALE. So the evaluation process of a particular ALE will not be easy.

2   Paradigm and Conceptual Changes in Education: New Foundation for the Design of Advanced Learning Environments?

It is important to situate our reflections in the context and beliefs of the author about learning technologies, new or old.

2.1 "All you need is Learning" (adapted from the Beatles)

The first idea is that technologies remain marginal in spite of their progress and their visible effectiveness in some fields of education. The learning process seems to be a side effect for the learner of an action of the teacher/educator or of an inter-action with a technology-based learning environment. This is particularly true in the traditional educational environment (the classroom, the University). The use of ALE seems to be more effective in educational situations far from this traditional conservative approach such as in distance teaching, people with special needs, vocational training, alternative schools ...

For us the future of research and design of ALE must be considered in the perspective of a radical change of education in the next years or decades.

All the ingredients of this change are present, especially in the western and developed countries, to provoque and facilitate the evolution, namely :

- The crisis of education and the anxiety of students and parents. It is common-place to say that the educational institution is in crisis and that it cannot satisfy the requirements and constraints of a modern world dominated by technology.
- Industrial and economic competition has stressed that a high level of education is required for most of the workers.
- The so-called "Information society" has put emphasis on information proces-sing and the required abilities and skills to survive in such a society.
- Most of the forecasts for the future society, sometimes qualified as post-capitalist, describe a society where knowledge will be the real richness and source of power. The emergence of a knowledge society is claimed not only by specialists of management and economy (Drucker 1993) but also by some philosophers.
- In parallel with the aforementioned crisis of education there has never been such a great celebration of learning. Learning seems to be the central paradigm of modern approaches or theories of management and organisation (i.e. the learning organisation).

Despite the occasional hype or fashion effects, there is an interest and a challenge for our future research because this evolution places learning (the learner?) at the centre.

2.2 Some Key Factors for a Change in Education

It is quite impossible to define here a clear view of the future of a transformed education, and even to give an exhaustive compilation of all the models, theories, pragmatic approaches which could characterize this future. We have only put emphasis on some directions, theories, ..., which could be of interest in the framework of the present work. This is just some pieces of the puzzle, without any "grand design". However the elements for change are selected because of their anticipated effects on research and design of future ALE. This is the same path that Pea (1992) follows in his investigation about Interactive Multimedia Technology for Learning.

There is an emergent corpus of theories, seminal writings, ..., which draws a potential future of advanced technology-based learning environments and new learning practices. Some elements a just listed here without any attempt to give a coherent construction, or even to exhaust their inter-relations or contradictions.

2.2.1 Some Myths or Paradoxes About Learning

The needs of a new conceptualization of learning requires a fresh view of education and learning, questioning our knowledge, beliefs, mental models about education. This radical shift is an obligation in order to reformulate, redefining the learning process. In a rather provocative reflexion about "locating learning in the times and spaces of teaching" G. Weade (1992) identifies three myths about learning:

- **Myth #1: Teaching causes learning.** In contrast, our assumption is that learning is a side-effect of teacher/instructor actions.
- **Myth #2: Learning is visible.** In fact learning is such a complex cognitive and social activity (intra-personal and inter-personal) that we cannot really observed it. This questions all the culture of assessments, measuring, grading that sometimes Computer Assisted Learning has enforced and systemized. But it questions also some approaches based on Actificial Intelligence, of which Intelligent Tutoring Systems are representatives.
- **Myth #3: Learning occurs in a "teachable moment".** In Wead's words: *"a "teachable moment" describes a point of in time when conditions seem ripe for students to make an association between relevant elements of a problem, or when they recognize a descrepancy between new evidence and the way they had expected to be".*

As a pragmatic result of questioning these myths, it can be said that learning is distributed in time and space with a distribution which is different from one individual to an another. In most cases only teaching can be located in time and space (for example in the classroom).

2.2.2 A Constructivist View of the Learning Process

This constructivist view is represented either by the cognitive approach initiated by the work of J. Piaget or the more social approach which take its roots from, for example, the work of L. Vygotsky (1962). The first one focuses more on individual (intra-personal) cognitive learning, whereas the second one focuses on inter-personal social construction of learning and on the important role of the social environments of the learner.

The impact of a "socially shared cognition" (Resnick, Levine & Teasley 1991) on design of ALE is less effective. But there is now a lot of works in this field, see for example works of Doise & Mugny (1981) or Perret-Clermont (1979).

From our point of view there are some consequences of considering these theories: we must always design and implemente ALEs which have an openess to support the potential mediation of the educational agents (teacher, tutor, parent, other learners). This is also an important background for the development of Computer Supported Collaborative Learning Environments.

2.2.3 A Situated View of Learning and Teaching

Another approach, complementary to the precedent, take its roots in the applications of linguistic, especially socio-linguistic, anthropology, sociology and ethnomethodology. Until now there have only been a few attempts to analyse classroom practices with methogologies derived from this field. However, it is over a decade that these fields of research, applied to computer and telecommunication technologies, have rapidly gained a great interest, especially in the field of information systems, HCI and Computer Supported Collaborative Work (CSCW). The role of language in the office has been analysed by some scientists such as T. Winograd & F. Flores (1986), and many others since this previous work. The anthropologist's viewpoint on human-machine communication was first applied by L. Suchman (1987) and has been the starting point of a new stream in the analysis and evaluation of HCI which borrows many ideas from ethnomethodology. From the concept of "situated actions", as emphasized by L. Suchman, the concept of "situated learning" and its related theory can be established. J. Lave & E. Wenger (1991) define this theoretical perspective as *"the basis of claims about relational character of knowledge and learning, about the negociated character of meaning, and about the concerned (engaged, dilemma-driven) nature of learning activity for the people involved"*.

2.2.4 A "Communicative" View of Learning and Teaching

An another perspective for the paradigm(s) of learning is the rejection of learning viewed as a transfer of knowledge. This follows the assumptions of constructivist theories, which state that knowledge is a personal and intra-personal construction. Many scientists and theorists have critiqued the "learning as transfer" paradigm with its associated transmission metaphor. But we must consider that it is the latent metaphor of the designer of most of the learning technologies and explains their fascination for television, cable networks, satellites ...

In a communication viewpoint, the relation between a learner and an educational agent for example, is not described only in transmission terms (the "Conduit-Metaphor"). Rather, each participant is active in the reciprocal construction of meaning where messages, utterances, text or conveyed through actions have no sense in theirself.

This communication model for knowledge has been recognised in the field of Artificial Intelligence and knowledge systems (Wenger 1987, Waërn et al. 1992). For example, Y. Waern et al. define different kinds of knowledge needed for communication and their implications for knowledge acquisition which is central in the domain of expert systems.

In the field of HCI and usability this leads us to examine the "repair" concept. In conversations, "repair" takes a great place. It is a frequent activity of the speakers for solving conversational breakdown (misunderstanding). As Brown and

Duguid explain, "repair" is a *"crucial feature in distinguishing design as process from design as product"*.

This is a central issue of the design of new technical artefacts. Because in the design as a product (the result is a piece of hardware, a software, a system) it is quite impossible to offer "communicative repair"; this forces the designers to try to design idiot-proof systems. Of course this expectation is in contradiction to the "situated actions" performed by the user. In the field of educational technology this has a resonance: many psychologists of learning or teachers know that learning is a process not a product. However the aims of educational technology, especially in computer-based learning, is often to transform the process into a product with different activities for the learner (i.e. separation between authoring and learning) with only few possibilities for repairing, and mostly without a knowledge of the situation of the learner and her social environment.

2.2.5   A Collaborative View of Learning

Cooperation and collaboration have always been present in the learning process. Some practices have even put cooperative learning in a central place (Johnson & Johnson 1991). This could be seen as natural in the framework of a socially shared cognition model or in a constructivist and situated learning approach. But, in fact, most of the traditional practices in the classroom or lecture situation in universities prevent real collaborative learning. Educational technology, especially computer-based learning, has even a more negative effect on collaborative learning, because it enforces the individualization of learning, and by design leads to an isolation of the learner.

From our point of view, there seems to have been a misunderstanding between the positive recognition of the plurality of cognitive or learning styles of different learners and individualization as a way of life induced by the "liberal society" ideology. The first part takes into account the possibility/theory of multiple intelligences and it can be named Individual-Centered Learning. If forecasts of technology of education keep to these general principles, instruction should be individualized (Nickerson 1988), it must not be opposed to the necessity of a socialisation of the learner and of the learning process.

There are another contributions which also prove the necessity of introducing collaborative learning. These are drawn from considerations about the present and future nature of work and organisations. For example, Bjorn-Andersen & Ginnerup have shown that enhanced communicative competence and cooperative problem-structuring capability will be the characteristic of both the CSCW systems and of the cognitive organisations that they support (Bjorn-Andersen, Ginnerup 1991).

From the management viewpoint, it is clear that Teamwork has never been so celebrated. Hence the "reflexive" claim: "Learning for teamwork, Teamwork for Learning" (Derycke 1993).

2.2.6   The Role of Practice in Learning

There is another source of reflection for change in education which is based on the role of practice (that of the learner) in construction and communication of knowledge. This approach is sometimes expressed by considering the learner as a self-practitioner or a self (personal) scientist. This questions of course the place and the role of laboratories, and workshops in education, and puts the accent on the "deductive vs inductive" debate in the knowledge acquisition.

But it must be also recalled that some abilities, skills, knowledge cannot be taught. In the taxonomy of pedagogical objectives this is the case of the higher levels such as expertise, evaluation,... It is true for the design abilities whatever the field considered: design of software, design of a house for an architect, design of a piece of music for a composer... In simple words: design can be learned but it cannot be taught.

Intuitively we know the central role of practice and of coaching for developing this design ability. In common practice in education, this is often solved by special case-studies solved in a group, or set-up of architectural studio, and mostly by integrating a period of real insertion of the student in a professional environment during his curriculum.

Again the first theoretical works in this field of Action Science, especially those of Schön (1987), have recently gained of lot of interest from a part of the HCI or Information System Design Communities. As a pragmatic result for us we can take care of integrating Action and Reflection in ALEs dedicating to learning of design (Fischer et al. 1992).

3   Present and Future Research in the HCI Domain

3.1 Some Clarifications About HCI: Design Space and Evolution

3.1.1 Design Space of Interfaces

The field of HCI is relatively new and the terminology and concepts are not always very clear. David Frohlich (1991) has proposed a common and systematic framework for describing the design spaces of interface. First of all, he separates the input interface (from human to computer) to the output interface (computer to human). His taxonomy is based on:

- *modes:* language and actions appear to be two fundamental modes;
- *styles: a recognized class of methods supporting interface activity.* In this category we find: command language, programming language, menu selection, form filling, icon, window and graphical interaction;

– *media: a representational model system for exchange.* In this category can fall audio, visual and haptic* information;
– *channels: an interface accross which there is a transformation of energy.* Roughly it is determined by the physical apparatus and human psycho-motor sensors.

Figure 2 gives one of the results of this approach. It is interesting to note that these representations are broad and exhibit some elements, for example haptic channels, which are not generally take into account explicity even if they are always buried in the system. For the educational field it is important to note that modern HCI research recognises the role of gesture in interactions. But there is a long way to follow to introduce this mode of interaction in ALE. Remember that in spite of rapid progress of Graphical User Interfaces and of the manifest power of images, our education is mostly text-centered (ie the Logocentric curriculum of M.A. White 1989). It appears that the design space is large, including multiple items, so future HCI would offer either some privileged paths in this space (profiles) or a plurality of styles, modes, medias, channels to a same application (development of multi-media/modal, ..., interfaces).

This kind of representation gives also a new look to the direct manipulation mode which has been popularized by interfaces "à la Macintosh".

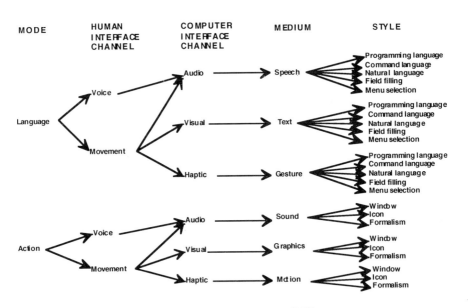

Figure 2. The input interface design space (Frohlich 1992)

_____

* In Frohlich's work "haptic" is used rather loosely to refer to both tactile information specifying surface contact and kinesthetic information specifying effector position.

From the cognitive viewpoint we know that Direct Manipulation has some roots in the cognitive constructivism pioneered by Piaget. This explain the success of this style of interface for computer-based learning environment which favour action and engagement of the learner (pro-active attitude versus reactive attitudes).

However the opposition of Direct Manipulation (Action modality) to the language modality, with severe critics of this late mode, must be tempered. The language mode must be also used and improved (Brennan 1990).

The last observation is the place of speech (both in input and in output) which indicate one of connections of this field of research in HCI and those of Artificial Intelligence.

3.2 Some Directions for Changes in the User Interface

Recently, there have had a lot of papers about the next generation of UI and research issues for the HCI field (Arens 1991), (CACM 1993), (Green & Jacob 1990), (Olson et al. 1993). It is too early to propose a synthesis of all these forecasts but we can exhibit the more salient points.

3.2.1 Some UI Shifts

Today User Interfaces must support a very wide range of applications or tasks and of users' profiles (from novices, casual, users, to professional experts...). For the next generation of UI some current limitations must be overcome. For example, A. Morse and G. Reynolds propose some strategy shifts in order to enable future UI to support new complex application domains (Morse & Reynolds 1993). Table 1 gives a summary of their proposals.

At the same time Y. Arens et al. (1991) define the next generation of UI's as Intelligent User Interfaces: *"User Interfaces that know what they're doing, User interface construction tools that know what they're building and better techniques for modeling existing media and the development of new media that match human communicative abilities better"*. The same authors define several future topics in:

- the field of user performance analysis;
- design models, exploration, and reusable modules;
- prototyping and optimization aids;
- help, documentation, and training;
- end user programming, tailoring and customization.

Whatever the ambition of this research program the different topics aforementioned could serve as a guideline to set up a specific agenda for research dedicated toward applications of progress of HCI to ALE.

Table 1. Summary of UI strategy shifts

Limitation	Strategic Shift
Point-and-click interfaces	Indirect manipulation Multiple isomorphic tools
WYSIWYS	You also need to see what is behind the curtain
GUIs are not graphical	Graphical modelling tools, packaged as reusable software components
Too many features	Rule-based design aids Demonstrational interfaces
Poor performance	Distributed processing Interoperability among applications
Government regulation	Distribute legal risk to suppliers

Excerpt from A. Morse and G. Reynolds (1993)

3.2.2 Major Issue of the Next Generation of UIs: Coping with Diversity

In order to cope with the diversity, of interaction styles and abilities of users, and plurality of cognitive styles (cf. Section 2) the future of UIs will be *multiple* (i.e., covering most of the design space).

Among the multi-features of the next generation of UI's we can consider:

– **Multimedia:** this is now a very hot issue for the HCI community with availability of very powerful hardware able to handle voice, moving video... A simple definition could be the harmonious integration of several media and supporting channels through the interface. The major issue is the support of temporal media such as sound (voice, music, noise) and video, with their afferent synchronisation problems at roughly three levels: at the hardware level (storage and network) at the document level (i.e., problems of hypermedia), at the interaction levels, especially in the field of Computer Supported Cooperative Work (videoconferencing for example).

The interest of Multimedia in the educational field is self-evident. For the first time it allows the integration of two traditionally opposed technologies for education: information technology and audio-visual technology.

Multimedia technology can improve the communicative aspect not only of Human-Computer Interaction by also of human-interaction; and associated with powerful digital networks it can give new communication modes with a better fit with human cognitive abilities (Felde 1992).

– **Multimodality:** A multimodal system is a system that supports many modalities for input and output (typing modality, graphic input modality via a mouse for example, speech modality). A modality can be seen as a privileged path through the input and output design space (Figure 2). Of course several modalities can coexist inside the same system.

Again this feature is in accordance with our general background for learning described in Section 2.

However, as W. Hill et al. (1992) pointed out:

— *"**Multimodality** per se is not enough. Why? Since tasks naturally span input modes in their own/peculiar ways, the blending of input modes is highly desirable. Blending of modes means at any point a user can continue input in a new, more pragmatically appropriate mode"*. This raises very difficult issues either at the design level or at the implementation level, with the need to adopt new software architectures or UIMS.

— **Multi-user:** In their survey of research directions for user interface tools Olsen et al stated that applications of advanced interfaces should, with appropriate extensions, permit multi-user interaction (Olson et al. 1993). This claims that all the applications software would be true multi-user systems even if the support of cooperative work is a special sub-domain with the design of CSCW systems, Groupware, ROOMWARE (supporting meeting room, real or virtual) and far beyond HCI (Olson 1993).

This is consistent with the assumptions made for the analysis of the background for the future of learning and its foundations (Section 2) in a communicative and situated view.

This is a way, not only to support intentional collaborative learning, but also to offer self-paced individualized computer-based learning environment which allows an educational agent to communicate with the learners in the context of their individual work. This could enrich the possibility of different kinds of tele-tutoring or offer some possibility of handling errors.

— **Multi-dimensionality:** The rapid growth of powerful hardware to support high resolution displays, true color and some operations relative to 3D representations (rendering, hidden part...) associated with advance in software, gives the possibility to represent data, information through the Graphical User Interface in multiple-dimension either realistic (3D) or virtual (more than 3D) (Benedikt 1992). This will be soon true also for sound restitution, thanks to the rapid progress of Digital Signal Processors: this already has hi-fi amplifiers for home-theatre which give a multi-dimensional restitution (spatialisation) of the sound.

— **Multi-facet:** We put under this category all the aims of HCI designers to give flexibility to their UIs. In most of the papers already cited there is a strong emphasis on features such as malleability, tailorability, customization, adaptable interface.

Roughly, we can separe these features in two classes:

— One, including tailorability and customization, is to cope with different classes of users (learners) and different context of use (community of practice). The use of these features is handled by the designer.

– The second class want to cope with individual diversity by providing malleable interfaces (possibility for the user to change some part of the interface) or auto adaptivity (the interface reacts to the user and alters its characteristics within the application environment). Many specialists in the field propose an "Intelligent User Interfaces" approach, ie which integrate progress of A.I. For example, Kühme et al. (1992) present a survey and a taxonomy for this kind of research activity with the information required for adaptation, the different strategies of adaptation and the underlying models and system architectures to support this adaptivity.

Of course these ways of research in the HCI fields have a link with the communicative viewpoint and the impossibility of designing idiot-proof applications/ products.

3.2.3 Evolution of the Design of UIs

In relation to the different conceptual changes in the field of HCI and the potential of new technologies, there is a need for an evolution of design methodologies and design tools. First of all, there is an evolution of the models used in the design of UI systems from the well-known Seeheim model to more sophisticated models (UIMS 1992).

Most of the previous models have in common the principle of separability of the user interface functionality from other functionalities (those of the applications) and the deriving of an architecture to support this separation. Even if this model has been valuable from the implementation viewpoint, we think that this separability "a priori" is questionable and in divergence with our previous assumptions about the interface concept. To try to reconcile the software requirements of dividing the systems into objects or components in order to facilitate maintenability and reusability, and the smooth transition principle, it is necessary to derive a new model which offers more different layers to offer a more gradual transition between interaction elements and application functionalities. The Arch model could be a good candidate for such a model. Figure 3 gives an overview of the model. This bridges the two domains, UI software and applications software. For advanced learning environment the domain-specific-components and the relative domain-adaptor-component will be probably the more differentiated far on the professional applications and strongly influenced by the underlying pedagogical design.

– The "Participatory design" (or collaborative design). This is a new framework for the design of complex socio-technical packages. It takes its roots from the Scandinavian approach relative to the introduction of new technologies in the workplace involving the users in the design processes. Participatory design is in accordance with some our general assumptions : better social insertion, situated action, community of practice... For a good survey of this "movement" you can see the special issue of communication of the ACM (Participatory 1993).

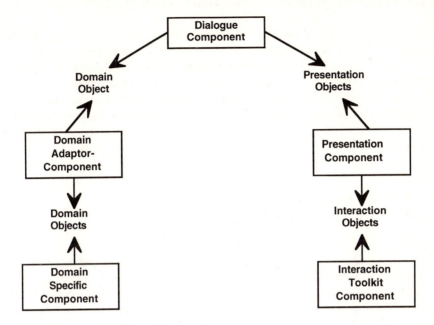

Figure 3. The interfaces between the components

Again it is fine to state that some of the scientists involve in the field of research share some common background with the world of education. For example, J. Greenbaum and M. King (1991) in their book "design by doing" explain that "Doing" was a central concept for us, embodying *the idea that there needs to be active involvement of users and designers working together at activities that actually do something"* and they write that it is founded by the work of famous educator John Dewey in "Learning by experience" (1938). They raise also an issue relative to "the cooperation between the practicing reflectioner and the reflecting practicioner" that could be linked to the role of practice (cf. Section 2.2.7).

At this time it seems that there is little participatory design of ALE. Of course, the major difficulty is the involvement of the learners. A first step in this direction will probably be to use participatory design with the participation of educators in the design process. But this introduces an another research issue: how to design methodologies and tools (for example graphical representation) which facilitate the participation between the ALE designer, the UI designer, the educational technologist and the representatives of educators.

3.2.4 Computer Supported Cooperative Work (CSCW)

We have already pointed out that some ideas, supports, etc., from the CSCW field can be grabbed to develop computer supported collaborative learning activities. This field of research emerged in the mid-1980s and covers different disciplines, from social sciences to computer sciences. Olson et al. (1993) present some the research issues for the 1990s for CSCW. By analogy, in the learning domain we can adapt some of these issues, for example, the need to understand the fundamental nature of the group activity, i.e., learning, that we are attempting to support.

4   A First Look at Research Issues at the Crossroads of HCI and ALE

The factors mentioned in Section 2, combined with the rapid evolution of computer and network technologies, entail a rapid evolution of technology-based learning environments and associated practices. From all the previous collected materials it is possible to draw first issues for future research which will be at the crossroads of the HCI research domain and of the ALE and Educational Science research domain.

4.1 The Metaphorical Issue

HCI is very influenced by the nature of the metaphors, latent or explicit, used either during the design process, or directly (visible) in the interface implementation (Sorensen 1992). In the field of modern uses of the computer, for example office automation, the desktop metaphor always associated with WIMP kind of interfaces, is now largely accepted. It must be recalled that this metaphor is centered primarily on the notion of documents and associated tasks. Of course the desktop metaphor is very questionable in the framework of applications to the learning process. Even in the field of office automation this metaphor, with its multi-window screens has been criticized because it introduces several drawbacks in the ergonomy due to the complexity of the task of mastering the interface. A new metaphor, at a meta-level, has been proposed by scientists of Xerox Parc and it is called the ROOMS metaphor: all the tools engaged in an activity are enclosed in a Room. To change the activity the user change of Room by clicking on the door (Robertson, Card & Mackinlay 1993). This room metaphor has been extended with benefit of rapid progress in 3D graphical devices. The Rooms have surfaces on which the data can be projected to form, for example, a true 3D spreadsheet. It is a very interesting direction of development because the ROOMS metaphor is consistent with other advanced fields of research in HCI: with hypermedia because navigation is central, with CSCW because a room can be a meeting room, and even with virtual reality.

In the same way, i.e., use of 3D graphical representation, some scientists have proposed new interface metaphors, alternative to the desktop metaphor, which integrate theoretical and experimental results about mental representations of tasks, procedures, and devices (Eberleh 1991).

Key Research Issues

From the ALE viewpoint the major issue lies in how to design UIs which take into account the change of metaphor and how to take advantage of advanced graphical hardware and software.

For example, in a Delta project called CO-LEARN, which is mostly dedicated to support cooperative learning at a distance, we have used the Room metaphor to build the interface as a way to access a virtual learning resource centre (Derycke 1993).

Other interface metaphors must be also considered :
– One class of metaphors is organized around the concept of agents following the evolution of artificial intelligence and the development of Distributed Artificial Intelligence. The interface agents metaphor is not without introducing problems in the human to computer interactions due to the anthropomorphizing of the interface (Laurel 1990).
– The second class postulates the need to have a "dramatization" of the interface. So the theatre becomes a powerful interface metaphor, giving new design principles for human-computer activity (Laurel 1991).

In spite of their interest, as fresh ideas to design advanced HCI, these approaches have not, until now, been put into concrete applications.

4.2 The Cooperative Issue

We put in this section the necessity to see the overall systems and particularly its UI as a cooperative system : the computer is a cooperative partner of the user (a collection of software agents) but it is also a tool, a medium to favour human to human cooperation.

The design of "cooperative software" is proposed to help people in quests for understanding (Rettig 1993). Fischer and his co-workers have done very interesting work in this direction.

Several important things must noted:
– First, because Fischer is interested by design of computer-assisted design environments, for example for architects, he grounded his work in the theories of Schön, among others (cf. the role of practice in Learning) (Fischer, Lemke & Morch 1991).
– The design rationale underlying the design process means explaining, deriving and justifying design decisions. This argumentation activity is often a cooperative

activity among members of design teams, and implies also a "communicative" view (cf. Section 2.2.3) (Fischer et al. 1992).

– The use of the critiquing approach to build knowledge-based interactive systems (Fischer & Girgenshon 1990). The critiquing paradigm is a way to present a reasoned opinion about a product or action in the framework of a problem-solving activity. Fischer also says that critiquing systems can be effective to support the learning process by offering to intelligent tutoring system an alternative to passive help systems.

The other side of the cooperative issue is the design of UI to support various kinds of collaborative learning activities: teleteaching, teletutoring, group problem solving, case-studies either in real-time using audio and video computer supported conferencing systems; or in deferred time (asynchronous) using computer mediated conferencing systems (Kaye 1992, Derycke 1993, Henri & Rigault 1996).

The "cooperative software" is also present in the field of research on Intelligent tutoring systems or knowledge-based learning environments with the attempt to change the established paradigm. It can be seen as a shift from the "socially shared cognition" theoretical stream to the socially distributed cognition one.

P. Dillenbourg and J. Self (1992) described this new field of computer-based learning and called it HCCL for Human-Computer Collaborative Learning. This is the first attempt to give a better socialisation flavour to knowledge-based learning environments.

4.3 The Multimedia and Hypermedia Issue

We have already stated that computer-based multimedia is very valuable for support the learning process because it can cope with the diversity of learning styles and the need for high interactivity (constructivist viewpoint).

We don't consider here the multimedia encyclopedia or other CD/ROM or CD/I based materials in spite of the interest for the education world. We want just to emphasize some of the potential problems and associated issues due to the design of UIs in the context of interaction with multimedia environments and hypermedia documents. There are already some advanced applications of multimedia which can prove the potential of multimedia technology inside UIs. For example, M. Brown et al. have demonstrated the power of graphic, color and sound in the animation of algorithms and then the usability for a programmer (or a student in computer sciences) to explore software program behaviour.

An another interesting direction of investigation is given by Feiner and his co-workers (Seligmann & Feiner 1989). In the IBIS (Intent-Based Illustration System) interactive multimedia is linked to a knowledge based system to automatically produced illustrations guided by communicative goals, following the way defined by Wenger (1987).

From the ALE viewpoint this raises several issues: for example what is the good use of sound especially voice in the UI of future ALEs? (multimodal issue), how to mix knowledge-based system or Intelligent tutoring system and the power of images ? (today the ITS are text-centered!).

For hypermedia the issues are relatively closed. If we envision here only hypermedia as an evolved UI technology, it is clear that major issues such as the navigation technique and tools, the semiotic and rhetoric of the anchors and links still remain in the learning domain of applications. But the specialisation of hypermedia for learning environments can give other issues which until now have received little attention (Dillon 1990, Jonassen 1990).

4.4 The Malleability or Flexibility Issue

Malleability, flexibility, tailoring, customizing are in the list of best wishes for the next generation of UIs. But what are their significiance in the framework of learning environments. One kind of flexibility, from the user viewpoint, can be achieved by adoption of multimodal interfaces where the user can use the most appropriate interaction devices and channels with which he feels confident or efficient.

The tailoring of the UIs (by the learner or the educator) is also very interesting to provide a better fit with the expected abilities, expertise,... of the learner and "culture" of the community in which practice will take place. However it could be achievable if tailoring tools will be powerful and simple to use by educators in the field.

We have shown that many designers plan to achieve malleability in the near future by selecting an UI architecture close to those of Distributed Artificial Intelligence. So the issue is: what is the specificity of the learning activity in this framework?

At a more general level this flexibility, which seems to be required for the next generation of professional software, does it is always require, in the education domain a permissive or prescriptive interface?

4.5 The Immersion Issue

We think that "Immersion" can be used to describe the next plateau in evolution of HCI technology. It appears immediately that the Immersion concept, in an artificial environment or world, has a strong relationship with a new field of research in HCI: Virtual Reality. The technical dimension of Virtual Reality is effectively a part of the interface technology with its helmets, data-gloves,... But Virtual Reality goes largely beyond the interface level and has an important cultural dimension and even its first cyberspace age philosophers. Of course the potential use of Virtual Reality raises some ethical issues and gives a new turn to the debate about people and technology (Beardon 1992).

From the learning viewpoint it seems that Virtual Reality is first an extension of the simulation technology already heavily used in training. This is the case of flight, tank, ship simulators for the training of the future pilots. In this perspective Virtual Reality can be seen as an improvement of the "reality" of the simulation and a way to produce simulators at less cost because most of the functions can be shifted from hardware to software.

But Virtual Reality can be also analysed in the light of some perspectives we have developed in Section 2:

– Virtual Reality is perhaps the extension of the microworld approach favoured by the cognitive constructivists.

– Virtual Reality has connections with "reflection in practice" and its implication for the learning process (Schön 1987). W. Isaacs and Senge (1992) take this theoretical perspection of action science in order to overcome limits of learning in a computer-based learning environment. In some domains where it is difficult to teach the subject, e.g., architectural design mentioned by Schön, or management and policy making mentioned by Isaacs, it is proposed that the learning process or loop has a phase of Immersion in a Virtual world: a constructed representation of the real world of practice. This can give a new start to the role and practice of case-studies in education and to design learning (Traub 1991). The "Virtual World" learning environment could be modelled and implemented by borrowing major advances in Virtual Reality research.

Key Research Issues

Probably the major issue is research on methodology for the design of these Virtual Worlds to support a learning process. The second is to construct learning tools to support the learner in his knowledge acquisition in order to avoid what it is called in the Hypermedia field the "Museum Syndrome": we have done a nice trip in the Virtual World with a very rich palette of sensations, images,... but we are not able to construct any deep knowledge from this experience.

4.6 The Design Issue

It has been shown that the design process of UI is still difficult, due to the lack of well accepted methodologies and of available usable tools. In the perspective of Advanced computer-based learning environments, this design issue is even more difficult because we have already said that is often difficult to involve the final user, the learner in the design process. However, if we maintain the Participatory Design principle, involving representatives of the learners, the main issue is probably to develop a set of methodologies and design languages (language used to design a product) which is understandable and efficient for the two: designer and user. For the analysis phase traditional methodologies to acquire the user's requirements are not valid because of the innovative characters of the product, advanced learning environments, which are under design. But some solutions have recently appeared: the scenario approach for example. Again this scenario approach is consistent with an "action-science" perspective and Participatory Design principles. For Carroll and Rosson (1992) each scenario is a description (in text, in a storyboard, etc) of the activities a user might engage in while pursuing a particular concern (ie in our case learning and/or teaching, tutoring, coaching). The scenario approach has also been advocated for the design of telecommunication applications for open and distance learning (Zorkoczy 1993).

4.7 People with Disabilities

We do take care to design the UI interface of some advanced learning environments for usability for people with special needs and disabilities. The "design for disability" in HCI has just begun to find an interest from HCI Community (for a survey see the ACM special issue "Computers and people with disabilities", 1992). This is not only a problem of ethic but a problem of efficiency, and even of market. Don't forget that a large segment of the population has some form of disability (for example they are color-blind, think of this for the design of color screens).

This is a difficult issue not only for its technical implications but also for its cultural dimensions: most of the designers of UIs have no knowledge of the impact of disabilities on usability and design experience.

However some of the new UI technologies, such as multimodality and multi-media, can offer part of the solution.

4.8 The "ClassroomWare" Issue

By "ClassroomWare" we want to put all the issues and problems relative to the insertion of the ALE, the technical artefact, in the learning place. The problem is broader than just learner interaction with computer, because now with the availability of videoprojectors, liquid crystal displays for overhead projectors, Local Area Networks, Videoconferencing systems, it is possible to set-up very sophisticated group learning environments. In the field of CSCW the science of the design of tools to support group work (or meeting) is called RoomWare.

In some cases this leads to the design of lecture rooms or theatres supported by networks and computers : the "electronic classroom". For example this has been done at Maryland University under the auspices of AT&T and it is call a "Teaching Theatre" (Norman 1990).

Key Research Issues

- Design of Electronic blackboard with some "intelligence" to support it, and an active interface.
- Improving the quality of sound (capture, restitution) in electronic classroom especially in the distance teaching set-up.

Conclusion

Availability of powerful new technologies for the computer interface such as multimedia, multi-modality, 3D graphics,... can give a new start to computer based learning environment which can cope with the diversity of "intelligences" of the learners, their plurality of learning styles and their social contexts. The design of advanced learning environments, using the power of information and communication technologies, would support a better social insertion of technologies for learning in order, not only to favour committement, and knowledge construction for the learner, but also to support cooperation and socialization in the learning process.

Acknowledgement

This article was partially derived from the experience we have gained in the framework of an EC DELTA project grant called CO-LEARN. I should like to acknowledge the stimulating collaboration of many members of our research teams: Laurent Barme, Pascal Croisy, Danièle Clément, Hervé Devos, Frédéric Hoogstoel, Claude Viéville. A special thank to Isabelle Logez and Anthony Kaye (Open University) for their help.

References

(NATO ASI Series F volumes are indicated by F and volume number)

Adler, P.S., Winograd, T.A. (eds.) (1992) Usability: turning technologies into tools. Oxford, UK: Oxford University Press

Arens, Y. et al. (1991) Intelligent user interfaces. Report ISI-RR-91-288. Information Sciences Institute, University of Southern California, Marina Del Rey, CA

Beardon, C. (1992) The ethics of virtual reality. Intelligent Tutoring Media 3(1), 23-28

Benedikt, M. (ed.) (1992) Cyberspace: first steps. MIT Press

Bjorn-Andersen, N., Ginnerup, L. (1991) The support of cognitive capacity in future organisations: towards enhanced communicative competence and cooperative problem-structuring capability. In: Rasmussen, J., Andersen, H.B., Bersen, N.O. (eds.) Human-Computer Interaction: research directions in cognitive science, European perspectives, Vol. 3, pp. 113-151. Hillsdale, NJ: Lawrence Erlbaum Associates

Brennan, S.E. (1990) Conversation as direct manipulation: An iconoclastic view. In (Laurel 1990), pp. 393-403

Brown, M.H., Hershberger, J. (1991) Color and sound in algorithm animation. IEEE workshop on visual languages, October 8-10, pp. 10-17

Brown, J.S. Duguid, P. (1992) Enacting design for the workplace. In (Adler & Winograd 1992), pp. 164-197

CACM (1992) Computers and people with disabilities. Communications of the ACM 35(5), 32-93

CACM (1993a) Graphical user interfaces: the next generation. Communications of the ACM 36(4), 36-109

CACM (1993b) Participatory design. Communications of the ACM 36(6), 24-103

Carroll, J.M., Rosson, M.B. (1992) Getting around the task-artifact cycle: how to make claims and design by scenario. ACM Transactions on Information Systems

Doise, W. Mugny, G. (1981) Le développement social de l'intelligence. Paris: Intereditions

Drucker, P. (1993) Post-capitalist society. Oxford, UK: Butterworth-Heinemann

Eberleh, E. (1991) Browsing cognitive task spaces instead of working on the desktop: an alternative metaphor. In: Bullinger, H.J. (ed.) Human aspects in computing: design and use of interactive systems and work with terminals, pp. 419-423. Amsterdam: Elsevier

Felde, N.I. (1992) Multiplexing Media. IEEE Communications Magazine, May, 90-98

Fischer, G., Girgensohn, A. (1990) End-user modifiability in design environments. Proceeding of CHI'90, ACM

Fischer, G., Lemke, A.C., Morch, A.I. (1991) Making argumentation serve design. Human-Computer Interaction 6, 393-419

Fischer, G. et al. (1992) Supporting indirect collaborative design with integrated knowledge-based design environments. Human-Computer Interaction 7, 281-314

Frohlich, D.M. (1991) The design space of interfaces. In: Kjelldahl, L. (ed.) Multimedia: Systems, interaction and applications, 1st Eurographics Workshop, Stockholm, Sweden, April 18-19, pp. 53-69. Berlin: Springer-Verlag

Green, M. Jacob, R. (1991) SIGGRAPH'90 workshop report: software architectures and metaphors for non-WIMP user interfaces. Computer Graphics 25(3), 229-235

Greenbaum, J., Kyng, M. (1991) Epilogue: Design by doing. In: Greenbaum, J., Kyng, M. (eds.) Design at work, pp. 269-279

Henri, F., Rigault, C. (1996) Collaborative distance learning and computer conferencing. In this volume

Hill, W., Wroblewski, D., McCandless, T., Cohen, R. (1992) Architectural qualities and principles for multimodal and multimedia interfaces. In Blattner, M., Dannenberg, R. (eds.) Multimedia Interface Design, pp. 311-318. ACM Press Frontier Series. Reading, MA: Addison-Wesley

Isaacs, W., Senger, P. (1992) Overcoming limits to learning in computer-based learning environments. European Journal of Operational Research 59, 183-196

Johnson, D.W., Johnson, R.T. (1991) Learning together and alone: cooperative, competitive and individualistic learning, 3rd edition. Boston, MA: Allyn and Bacon

Jonassen, D.H., Mandl, H. (eds.) (1990) Designing hypermedia for learning. F67

Kaye, A.R. (ed.) (1992) Collaborative learning through computer conferencing. F90

Kommers, P.A., Jonassen, D.H., Mayes, J.T. (eds.) (1992) Cognitive tools for learning. F81

Kühme, T., Dieterich, H., Malinowski, U., Schneider-Hufschmidt, M. (1992) Approaches to adaptivity in user interface technology: survey and taxonomy. In: Larson, J., Unger, C. (eds.) Engineering for human-computer interaction, pp. 225-250. Amsterdam: Elsevier

Kuutti, K., Bannon, L.J. (1993) Searching for unity among diversity exploring the "interface" concept. Proceeding of INTERCHI'93 (ACM, IFIP), Amsterdam, April 24-29, pp. 263-268

Laurel, B. (1990) Interface agents: metaphors with character. In: Laurel, B. (ed.) The art of human computer interface design, pp. 355-365. Reading, MA: Addison-Wesley

Laurel, B. (1991) Computer as theatre. Reading, MA: Addison-Wesley

Lave, J., Wenger, E. (1991) Situated learning: Legitimate peripheral participation. Cambridge, UK: Cambridge University Press

McLean, R.S. (1989) Megatrends in computing and educational software development. Education & Computing 5, 55-60

Morse, A., Reynolds, G. (1993) Overcoming currents growth limits in UI development. In (CACM 93), pp. 73-81

Nickerson, R.S. (1988) Technology in education in 2020: Thinking about the not-distant future. In: Nickerson, R., Zodhiates, P. (eds.) Technology in education: Looking toward 2020, pp. 1-24. Hillsdale, NJ: Lawrence Erlbaum Associates

Norman, K.L. (1990) The electronic teaching theater: interactive hypermedia and mental models of the classroom. Current psychology: Research & Reviews 9(2), 141-161

Olsen, D.R., et al. (1993) Research directions for user interface software tools. Behaviour & Information Technology 12(2), 80-97

Olson, J.S., et al. (1993) Computer-supported co-operative work: research issues for the 90s. Behaviour & Information Technology 12(2), 115-129

Pea, R.D., Gomez, L.M. (1992) Distributed multimedia learning environment: why and how? Interactive Learning Environments 2(2), 73-109

Perret-Clermont, A.N. (1979) La construction de l'intelligence dans l'interaction sociale. Berne: Peter Lang (see also Perret-Clermont, pp. 41-61 in Resnick 1991)

Resnick, L.B., Levine, J.M., Teasley, S.D. (eds.) (1991) Perspectives on socially shared cognition. Washington, D.C.: American Psychological Association

Rettig, M. (1993) Cooperative Software. In (CACM 1993a), pp. 23-28

Robertson, G.G., Card, S.K., Mackinlay, J.D. (1993) Information visualization using 3D interactive animation. In (CACM 1993a), pp. 57-71

Schön, D.A. (1987) Educating the reflective practicioner. San Francisco, CA: Jossey Boss

Seligmann, D.D., Feiner, S. (1989) Specifying composite illustrations with communicative goals, ACM CHI Conference, pp. 1-9

Sorensen, E.K. (1992) Metaphors and the design of the human interface. In F90, pp. 189-200

Suchman, L. (1987) Plans and situated actions: the problem of human-machine communication. Cambridge, UK: Cambridge University Press

Traub, D.C. (1991) Simulated world as classroom: the potential for designed learning with virtual environments. In Helsel, S.K. Roth, J.P. (eds.) Virtual reality, theory, practice, and promise, pp. 111-121. Ferry Lane West, CT: Meckler

UIMS (1992) A Metamodel for the runtime architecture of an interactive system. The UIMS tool developers workshop. ACM SIGCHI Bulletin 24(1), 32-37

Vygotsky, L.S. (1962) Thought and language. Cambridge, MA: MIT Press

Waern, Y., Hägglund, S., Löwgren, J., Rankin, I., Sokolnicki, T., Steinemann, A. (1992) Communicative knowledge for knowledge communication. Int. J. Man-Machine Studies 37, 215-239

Wead, G. (1992) Locating learning in the times and spaces of teaching. In: Marshall, H.H. (ed.) Redefining student learning, pp. 87-118. Norwood, NJ: Ablex

Wenger, E. (1987) Artificial intelligence and tutoring systems. Los Altos, CA: Morgan Kaufmann

Winograd, T. Flores, F. (1987) Understanding Computer and Cognition. Norwood, NJ: Ablex

White, M.A. (1989) Current trends in education and technology as signs to the future. Education & Computing, 5, 3-10

Zorkoczy, P. (1993) Educational scenarios for telecommunication applications. In F133

6

New Technologies for Modelling Science and Mathematics

Rosamund Sutherland

Institute of Education, University of London

Abstract: Within this chapter I discuss the ways in which current computer environments allow for the type of immediate behavior which is advocated by supporters of situated cognition. I discuss the tension between informal and formal approaches to modelling in science and mathematics and how this relates to the type of objects which students interact with at the interface. I suggest that the most difficult aspect of modelling is to be able to shift between thinking with the physical and the more abstract mathematical objects of a model. Research is needed on how new technologies support students to make this shift.

Keywords: Algebraic approaches, computer, informal knowledge, intermediate objects, microworlds, new technologies, situated cognition, spreadsheets, symbolic processing, variables

1 Introduction

Ten years ago when I started research on the use of computers for teaching and learning there were few computers available in schools and there was relatively restricted software for use in the classroom. This situation has changed dramatically but we still know very little about how to use the increasingly rich software-environments in the classroom. Most research and development in this domain focuses on the development of the software itself or on evaluation with small groups of students and thus overlooks the ways in which the learning environment may adapt to these new technologies. As Solomon has pointed out research in this area has an almost exclusive focus on the individual and he gives three reasons for this.

First, many of the researchers involved in the design, implementation and study of new technology-intensive instructional projects, being in fact involved in aspects of applied cognitive psychology, adopt psychology's emphasis on the individual,... Second, despite interesting attempts to conceptually describe classrooms as learning environments (e.g. [7]), a viable theory of such is yet to emerge to allow simultaneous study of individual and environmental changes within the same conceptual framework. Third, even if a conceptual framework for the study of individuals and learning environments would be available, no methodology has been proposed so far to make such a study empirically possible [20].

There is however (at least in the USA) increasing pressure for classroom validation of software. This emphasis on the classroom is having the effect of narrowing the gap between the more expressive microworld type software and intelligent tutoring systems "there has been a recognizable movement away from ITSs, curiously enough toward other forms of software that may or may not incorporate intelligence [11, 12, 19] – this trend appears to be a response, in part, to the difficulty of building realistic ITSs coupled with the growing push to create systems that are demonstrably effective" ([9] p. 241). Software development is tending towards more interactive and visually-based software (for example Cabri-géomètre [10]) and we are beginning to see the integration of ITS with more exploratory microworlds (for example ANGLE and Geometer's Sketchpad [9]). But whatever the type of software being used in the classroom Balacheff points out that "students will always try to find in the behavior of the educational software indications of their teacher's expectations, and then try to complete the task without necessarily functioning within the intended epistemological register" [1]. So this means that whether students are engaging with a more exploratory microworld or a more directed ITS they can avoid constructing the intended knowledge. This is often overlooked by those who advocate computer-based learning environments, primarily because of a lack of research in the teaching-learning setting.

2   The Debate Between Symbolic Processing and Situated Cognition

The current debate between advocates of the more traditional cognitive science *symbolic processing* approach with its focus on the individual and the internal and the more recent *situated cognition* approach with its focus on the external and the social is very relevant to the development of computer-based environments in science and mathematics. As Norman has pointed out "the proponents of each method tend to act as if there were two precise, exclusive philosophies of approach, one correct and the other wrong … the caricature of the traditional information studies of symbolic processing is that they focus entirely upon the processing structures of the brain and the symbolic representation … the caricature of the situated action approach to cognition is that these studies focus entirely upon the structures of the world and how they constrain and guide human behavior" ([16] p. 3, 4).

The cognitive science and the situated-action approaches appear to belong to two non-overlapping paradigms, despite attempts at integration. The theoretical objects and thus the objects of enquiry are different. So for example in the symbolic approach memory is conceived of as stored rules or schema structured in a representation language. In a situated cognition approach memory is conceived of as neural nets which are reactivated and recomposed during activity [5]. Clancey claims that the symbolic approach has developed a symbolic model of mind which is just a model, but which is used as if this is how the mind works. The symbolic processing approach has tended, at least in the past, to make inappropriate links between the external representations which we find in the environments and the internal representations of the mind.

Clancey maintains that the symbolic processing view has artificially separated theoretical objects and he proposes that "to be perceiving the world is to be acting in it – not in a linear input-output relation (act-observe-change) – but dialectically, so that what I am perceiving and how I am moving codetermine each other" (ibid., p. 95). For Clancey all activity involves construction by the individual and is not a question of rote retrieval of 'known facts'. Whereas the symbolic approach reduces comprehension to matching a stored body of memorized knowledge. Clancey also emphasizes that neurological structures are always new (not literally the same physical structure) and they are always activated as part of an ongoing coordination as circuits" (ibid., p. 103).

Much of this debate centers around the idea of *planning* and *planning in action*. The symbolic processing point of view suggests that plans, which are somehow stored in memory, influence action. Whereas the situated-action approach suggests that "every act of deliberation occurs as an immediate behavior. That is, every act of speaking, every motion of the pen, each gesture, turn of the head, or any idea at all is produced by the cognitive architecture a matter of course, as a new neurological coordination" (ibid., p.112). The differences in viewpoint are so extreme that they suggest different perspectives about 'being in the world', between those who want to maintain a-priori 'rationality' at all costs and those who accept that reasoning "occurs in sequences of behavior over time" (ibid.).

An emphasis on formal planning has always been questioned by Papert "in our culture, the structured, plan-oriented, abstract thinkers don't only share a style but constitute an epistemological elite" ([24] p. 369). The argument put forward is that recent technological developments are allowing for a more interactive way of working at the computer which is undermining the elitist 'hard planning' position (ibid., p. 371). This trend is linked to emergent AI theories with their emphasis on systems or societies of negotiating agents, in contrast to the earlier more rule-based AI theories. Papert and Turkle draw our attention to the fact that 'diving into' situations rather than looking at them from a distance, that connectedness rather than separation, are powerful means of gaining understanding" (ibid., p. 382).

The change in the nature of computer environments now makes it more possible to work in this interactive way in which representations are used to create and interpret our activities. This was much more difficult in the early days of computing in which reasoning (writing a program) was artificially separated from action (executing a program). So, in this sense, current computer environments now allow the more *immediate* behavior which has always been possible in paper-based environments. The 'hard' cognitive science theories with their emphasis on formal structure and symbolic representation have in the past not taken account of the immediate nature of problem solving although there does appear to be some move in this direction "Rather than assuming that competence in proof is derived purely from internalizing the formal domain rules, the cognitive model underlying ANGLE claims that proof knowledge is induced from and organized around prototypical perceptual configurations" ([9] p. 242).

3   Informal Versus Formal Knowledge

The evidence from the research of Lave and others showing that individuals solve problems in non-school settings very effectively and without recourse to formal taught school methods is not adequately taken account of by the cognitive science symbolic processing perspective. It is now widely recognized that students use a range of strategies to solve problems. This is the case for street sellers in Brazil [4], supermarket shoppers [13], and computer science undergraduates [8]. These strategies are just as much related to the needs of the situation as to the knowledge of the student. If we consider the following problem –

The Rectangular Field Problem: The perimeter of a field measures 102 meters. The length of the field is twice as much as the width of the field. How much does the length of the field measure?

– students could use the following (or a similar) formal analytic algebraic approach to solve this problem:

$$
\begin{aligned}
\textit{Let the width of the field} \quad &= \quad X \text{ meters} \\
\textit{Let the length of the field} \quad &= \quad L \text{ meters} \\
\textit{Then} \qquad\qquad\qquad L \quad &= \quad 2X \qquad\qquad (1) \\
\textit{and} \qquad\qquad 2L + 2X \quad &= \quad 102 \qquad\qquad (2) \\
\textit{So by substituting } (1) \textit{ in } (2) \\
4X + 2X \quad &= \quad 102 \\
6X \quad &= \quad 102 \\
X \quad &= \quad 17 \text{ meters} \\
\textit{So width of field} \quad &= \quad 17 \text{ meters} \\
\textit{Length of field} \quad &= \quad 34 \text{ meters}
\end{aligned}
$$

But they could also use a 'trial and refinement' approach as illustrated by the following solution: "well I tried 40, it was 120 ... so I knew it must be smaller than that ... in the 30s ... and when I tried 36 and it was 108 ... I knew it couldn't be 35 so it must be 34..."

Alternatively they might use a 'whole/parts' approach [14]: "I did 102 divided by 6 ... I just did two of the lengths to make it sensible ... I just thought there must be two of those in one length ..."

These last two more informal approaches differ in a number of respects from the formal algebraic approach. They involve manipulating the quantities presented within the problem. In the 'trial and refinement' approach this involves working with quantities until the given total length of the field is produced. In the 'whole/ parts' approach this involves working from the known whole (102 meters) and dividing this by the number of parts (6). In contrast to this the analytic algebraic approach involves working from the unknown (called for example x) to the givens within a problem. In this sense the algebraic approach is in opposition to approaches which involve thinking with the given quantities within a problem situation.

Working with students from different age groups we have found that many of them use non-algebraic approaches to solve problems which are similar to the 'rectangular field' problem [21]. Within the earlier cognitive science research the assumption was often made that an individual should represent the solution to a problem such as the 'rectangular field' problem with a structure in the mind which somehow matched the structure of the algebraic solution to the problem and computer environments were designed to provoke the 'correct' match. But if many students use approaches which differ substantially from the algebraic approach and which involve manipulating the quantitative objects of the problem, then they are likely to find it very difficult to learn the 'taught' algebraic approach whether being taught by a computer or a teacher. Taking into consideration the wealth of strategies used by students when solving problems implies a situated cognition approach to learning. The issue of how to teach the more formal algebraic approach is still problematic. Currently there is somewhat of a polarization in terms of teaching approaches. Either informal strategies are ignored and the formal is imposed. Or the informal is celebrated as an effective problem-solving strategy and the formal is de-emphasised. Taking into account students' informal strategies whilst teaching the more formal approach is the most difficult route.

We have had some success with using spreadsheets to help students move from a non-algebraic to an algebraic problem solving approach [21]. The symbolic spreadsheet language becomes a mediator in the process of both symbolizing the quantities within a problem and in moving towards thinking with abstract objects. Ultimately we want students to be able to shift flexibly between thinking with different types of objects, and we need to know which objects and relationships between objects a particular computer environment is favouring. The nature of the appropriate 'object to think with' is likely to be different in science than in mathematics. Feurzeig distinguishes between two types of models which can be represented by the computer, those that describe the phenomena mathematically and those that describe the phenomena in ways which express the behavior of the modeled objects and their interactions[6]. He suggests that "First it is likely that objects and object interactions often correspond more closely than the differential equations to the mental models in the scientist's mind. Though scientists converse comfortably about models in symbolic mathematical language they often envisage the model process visually and their discussions make extensive use of diagrams. Second and more important, high school students relate to simulations a great deal better than they do to differential equations' [6]. So what will be the role of mathematical modeling in the future, and how important will it be for students to be able to make links and shift between a mathematical model and a model of a physical situation? In an ongoing project we are investigating the ways in which students use mathematical modeling while solving problems in science (biology, chemistry and physics) and the ways in which the use of a spreadsheet facilitates this modeling process [21]. A spreadsheet enables complex phenomena to be modeled without using differential equations although it does not allow students to interact with a simulation of physical objects.

Developments in technology have made it easier to represent pictures of physical objects on the screen and this together with the evidence that students often think with these more realistic objects as opposed to the more abstract algebraic objects has led to a proliferation of computer environments in which students can somehow interact with images of physical objects. One development in the case of algebra word problems (still the focus of much cognitive science research) is to present students with pictorial representations of the objects of the problem situation [15]. Again, as with the earlier cognitive science research, the assumption seems to be that there will be a smooth transition between thinking with the 'realistic' objects and developing a more formal model using algebra. I conjecture that presenting students with these images of 'realistic' objects is more likely to provoke a non-algebraic approach to solving the problem as opposed to the algebraic approach which the software is trying to teach.

Another approach is used within the Microcomputer Based Laboratory (MBL) tools in which students interact with both physical non computational objects and formal representations of these physical objects [22]. Students use MBL tools to collect physical data which are then graphed in real time. These environments seem to offer considerable potential but as far as I know there has been little research investigating the ways in which students move between thinking with the physical non-computational objects and thinking with the more symbolic computational objects in these types of environments.

4 A Middle Way

Barbara White argues against what she calls the 'top-down approach' of presenting abstract algebraic formulae to pupils. This is the traditional approach in science which asks students to match the variables in a problem to the appropriate equation and then to manipulate this equation to arrive at a solution to the problem, because this approach obscures the underlying causal principles. She also argues against what she calls a 'bottom-up approach' advocated by situation cognition theorists in which knowledge is to be gradually induced from real-world experience. "I propose a middle-out approach in which students are introduced to new domains via causal models represented at an intermediate level of abstraction" ([25] p. 26). She advocates an approach to science and engineering in which students work with microworlds in which they interact with intermediate levels of abstraction. She gives an example from the ThinkerTools Project in which students interact with a dot-impulse model. The claim is that this intermediate abstraction helps middle-school students understand force and motion. Students control the motion of the large dot in order to make it navigate the track and stop on the cross. Other intermediate representations used are small dots which represent the history of the large dot's motion and the datacross which represents the velocity components of the large dot. The idea is that the student thinks of the large dot as a generic object (which could be a spaceship or a billiard ball). The student controls the motion of the large dot by means of a joystick and the task is to determine the physical laws underlying the dot's behavior. White claims that these intermediate causal models are understandable because they build on intuitive notions of causality and mechanism. The claim is also made that these generic representations facilitate the transfer to multiple contexts. The intermediate models employ both visual and verbal representations.

In order to achieve coherent, transferable expertise one needs to acquire a set of linked hierarchical models (such as the particle model, the transport model and a steady state model) in which the emergent properties at one level become the primitive properties of the next level, and which thereby ground the higher order abstractions (such as Ohm's law) with a mechanistic unpacking of the physical phenomena down to the particle level. What is needed, we conclude, is a learning environment in which these multiple, mutually-consistent conceptualizations of electricity can be learned [26].

White suggests that professional scientists also work with models at intermediate levels of abstraction when solving problems and when developing new theories. The intermediate causal model approach links into the intuitive scientific thinking of students and supports them to make links between intuitive models and more formal mathematical models. The emphasis is on integration of different models at different levels of abstraction.

5. Some Concluding Remarks

Clearly in any computer-based environment the type of objects which students interact with at the interface are crucial to their developing understanding of mathematical or scientific ideas [23]. These could be the more realistic objects of

Nathan's microworld, geometric objects of Cabri-géomètre, intermediate objects of White's microworld or the cells of a spreadsheet. The nature of computational objects is more often driven by developments in technology than by an awareness of the meanings which students construct from their interaction with these objects. More research is needed on the meanings which students develop when interacting with the rich and prolific environments which are being developed for science and mathematics. We need to know more about potential tensions between more spontaneous and more formal approaches and the role of teaching in this area.

Pea suggests that computational tools should become an extension of the student's own resources [17]. This is similar to the ideas put forward by Vygotsky in which the object of study becomes not the individual but the individual acting with mediational means, often called psychological tools. Examples of psychological tools are natural language, algebra, icons, graphs. The computer is increasing the range of psychological tools which can be made available to students. We need to understand more about the ways in which these psychological tools support students in the problem solving process and the ways in which students can make use of computer-based psychological tools in non-computer settings. We also need to know more about the ways in which the teacher can help students use these new tools to support the learning of mathematical and scientific ideas. This is where more research in the classroom is necessary in order to tease out the relevant issues and to influence ongoing technological developments. Here the work of Brousseau [3] and other groups in France can point the way towards a possible methodology which involves the simultaneous study of individuals and the whole group within their learning environment. As a starting point for this classroom based research we can predict what learning is likely to occur when students interact with a computer-based environment by analyzing the epistemological and didactical domains of validity of the software [1]. This analysis must take into account the dialectic relationship between external representation of knowledge and the knowledge being represented.

References

(NATO ASI Series F volumes are indicated by F and volume number)

1. Balacheff, N., Sutherland, R. (1994) Epistemological domain of validity of microworlds: The case of Logo and Cabri-géomètre. In: Lewis, R. (ed.) Lessons from learning, IFIP Conference TC3 WG3.3. North-Holland
2. Balacheff, N. (1993) Artificial intelligence and real teaching. In F121
3. Brousseau, G. (1986) Fondements et méthodes de la didactique des mathématiques, Recherches en didactique des mathématiques 7(2), 33-115
4. Carraher, D., Carraher, T., Schlieman, A. (1985) Mathematics in the streets and in the schools. British Journal of Developmental Psychology 3, 21-29

5. Clancey, W.J. (1993) Situated action: a neuropsychological interpretation response to Vera and Simon. Cognitive Science 17, 87-116

6. Feurzig, W. (1994). Visualisation in educational computer modelling

7. Hall, R., Kibler, D., Wenger, E., Truxaw, C. (1989) Exploring the episodic structure of algebra story problem solving. Cognition and Instruction 6(3), 223-283

8.

9. Koedinger, K.R., Anderson, J.R. (1993) Effective use of intelligent software in high school math classrooms

10. Laborde, J.M. (1996) Intelligent microworlds and learning environments. In F117

11. Lajoie, S.P., Derry, S.J. (1993) Computers as cognitive tools. Hillsdale, NJ: Lawrence Erlbaum Associates

12. Larkin, J.H., Chabay, R.W. (1992) Computer-assisted instruction and intelligent tutoring systems: Shared goals and complementary approaches. Hillsdale, NJ: Lawrence Erlbaum Associates

13. Lave, J. (1988) Cognition in practice. Cambridge, UK: Cambridge University Press

14. Linns, R.C. (1992) Algebraic and non-algebraic algebra. Proc. 17th Psychology of Mathematics Education Conference, Durham, USA

15. Nathan, M., Resnick, L. (1994) Less can be more: The potential of unintelligent tutoring based on psychological theories and experimentation

16. Norman, D. (1993) Cognition in the head and the world: an introduction to the Special Issue on Situated Action. Cognitive Science 17, 1-6

17. Pea, R. (1992) Augmenting the discourse of learning with computer-based learning environments. In F84

18. Rojano, T., Sutherland, R. (1993) Mexican/British project on the role of spreadsheets within school-based mathematical practices. Project funded by the Spencer Foundation

19. Schank, R.C. (1991) Where's the AI? Artificial Intelligence magazine 12(4), 38-49

20. Solomon, G. (1992). Differences in patterns: Studying computer enhanced learning environments. Psychological and Educational Foundations of Technology-Based Learning Environments, Crete, July

21. Sutherland, R., Rojano, T. (1993) A spreadsheet approach to solving algebra problems, Journal of Mathematical Behaviour 12(4), 351-383

22. Thornton, R. (1992) Enhancing and evaluating students' learning of motion concepts. In F86

23. Tiberghien, A. (1992) Analysis of interfaces from the points of view of epistemology and didactics. In F86

24. Turkle, S., Papert, S. (1990) Epistemological pluralism: Styles and voices within the computer culture. Signs 16(1), 345-377

25. White, B. (1993) Intermediate abstractions and causal models: A Microworld-based approach to science education. In Bona, P., Ohlsson, S., Pain, H. (eds.) Proc. Artificial Intelligence in Education Conference, AICE, USA

26. White, B., Frederikson, J., Spoehr, K. (1993) Reductionist conceptual models and the acquisition of electrical exertise. Proc. 15th Annual Meeting of the Cognitive Science Society. Hillsdale, NJ: Lawrence Erlbaum Associates

7

Implementation of Technology Education in New York State

Michael Hacker

New York State Education Department, 98 Washington Avenue
Albany, New York 12234, USA

Abstract: In New York, a new school discipline is emerging. Technology Education is taking its place alongside longstanding Mathematics and Science programs. Technology Education has the potential to actively engage students in design-and-construct activities that help them to synthesize the knowledge and skills in math, science, social science, and other subjects. As an emerging subject area in the school curriculum, Technology Education faces several issues that impede implementation. These issues relate to its roots in the craft tradition, its relatively recent evolution, and the present lack of public awareness and support. Current trends in New York focus on the use of Technology Education as a vehicle through which integrated Mathematics, Science, and Technology (MST) programs can be implemented.

Keywords: Assessment, curriculum, pre-college education, staff development, technology education

1   Introduction

One significant challenge for education is to provide a core of technologically creative and competent individuals who can contribute to human-centered technological development; that is, without unacceptable damage to humans or to the natural environment. It is the human who conceives what should be developed and it is the human who should control its destiny.

Just as society advances technologically, so must its educational system compensate and adjust for the changing world. The human dependence on technical means for survival and quality of life warrants the study of technology by all people.

Technology Education is an emerging discipline in the United States and internationally. Technology Education has sprung from the roots of Industrial Arts

Education in the United States, and Craft teaching in other countries. In New York, a massive inservice program has trained thousands of Industrial Arts teachers to deliver Technology Education. Forty states and sixteen foreign countries now have Technology Education programs in their public schools [1]. In England, Wales, and Northern Ireland for example, Technology as one of the foundation subjects in the new National Curriculum, is required of all students ages 5-16 [2]. In The Netherlands, Technology Education is required of all students as of 1993.

By design, Technology Education complements mathematics and science education. It is an integrating discipline whose programs provide students a unique opportunity to synthesize much of their learning. Real-world problem-solving provides the context for conceptual learning to be applied, and for technological design activity to occur.

2   Purposes of Technology Education

Technology Education programs assist students of all academic levels to discover their technical interests and capabilities; prepare for further education leading to careers in engineering, technology, and the skilled trades; and attain a share of technological literacy and capability.

The programs fulfill several broad purposes [3], which enable students to:

1) Use a systematic design process to create technological solutions to problems.

2) Apply knowledge and skills relative to systems, processes, and resources of technology to the creation of products, services, systems, and environments.

3) Recognize, analyze, make informed judgments, and take personal or group action regarding socio-technological issues.

4) Make informed decisions and choices leading to the selection, effective use, and maintenance of technological products, services, and systems.

5) Apply technological skills and knowledge to anticipated needs and opportunities in future employment, life long education, and personal fulfillment.

6) Synthesize knowledge and derive additional sense of purpose from the study of other disciplines, particularly mathematics, science, and social science.

3   School-Based Technology Programs

The driving force behind the emergence of Technology Education is the recognition that technological literacy has become a new priority for education. A recent NATO conference comprising academicians, governmental representatives, and industrialists from fourteen countries, recommended "the establishment in all

NATO countries, for all students throughout their compulsory schooling, of Technology Education programs [4].

UNESCO suggests that "every person in all societies must have the advantages of a basic education which includes scientific and technological literacy," but, in reporting the results of a recent global survey, indicates that:
Technology Education is certainly the most challenging and demanding of the "tri-part" science, mathematics, and technology education plan to be incorporated into basic education. There is no precedence in school for teaching it and, as a concept, it is little understood [5].

New York State is providing recognized leadership to this international movement. "Nowhere is the movement more advanced than in New York," [6] where Technology Education is a required component of all students' fundamental education. At the middle school level, students must study Technology for one year. At the high school level, sequences in Technology Education fulfill Regents and local graduation requirements. A complement of over 30 elective courses is offered.

3.1 Elementary Technology Education

In New York, some schools offer Technology Education at the Elementary School level, although there is no State mandate requiring this program. Certified Technology Education teachers deliver the instruction in some cases; in other instances, common branch teachers, resourced by a subject matter specialist, integrate hands-on Technology Education into their teaching. A State syllabus is available for Elementary School Technology Education.

3.2 Middle Level Technology Education

Science and Mathematics are disciplines with longstanding educational traditions in New York. As of 1987, the Board of Regents has required that all students in the middle schools also take a minimum of one-unit of Technology Education by the completion of eighth grade.

Since technology is by its nature an integrating endeavor, the Technology Education program reinforces and gives additional purpose to mathematics, science, and social studies study by applying such learning to the design and construction of technological solutions to problems. The program focuses on a study of technological tools, resources, processes, and systems, and the impacts on people, society, and the environment.

A distinction must be drawn between Technology Education, and Vocational/ Technical education programs which are intended to provide students with entry level career skills. Technology Education is a component of all students' fundamental education. Its mission is to promote general technological literacy and capability. It thereby serves to prepare students for citizenship in a technological society, as well as for further education leading to technological careers.

The following is an overview of the required one-year Technology Education requirement that is in place for all New York State middle school students [7].

3.3 Conceptual Curriculum Framework

The engineering systems model (input, process, output, feedback and control) provides the conceptual curriculum model for the New York Technology Education program. The one-year program is divided into ten modules, five in grade seven, and five in grade eight.

At the seventh grade level, the Nature and Evolution of Technology; System Inputs (defined as the desired results and the resources used by the system); System Outputs (intended and unintended results); and Technological Problem Solving provide the content. The Eighth Grade Program "closes the loop" in our systems-based technology model, and adds concepts of processing, feedback, and control.

Several contexts are suggested from within which examples might be drawn. These are systems in Biotechnology, Communications and Information Technology, Production Technology, and Transportation Technology. An overview of the middle school program follows.

3.4 Seventh Grade Program

Module 1 introduces the student to this new field of study, addresses the evolution and importance of technology, and focuses on reasons why people must study about technology.

Module 2 focuses on the seven resources which are generic to all technologies. These are people, information, tools and machines, materials, energy, capital, and time.

Module 3 is a problem solving module where students could be challenged to solve a prescribed technical problem, design a device for a handicapped person, or participate in a competitive problem solving event. It serves as a natural introduction to the systems module.

Module 4 introduces students to systems terminology. Inputs, process, and outputs are studied. The concept of feedback is lightly touched.

Module 5 focuses on the positive and negative impacts of technology (the outputs) as well as the match between technology and the environment, and technology and the human user.

3.5 Eighth Grade Program

Module 6 revisits the resources for technology, however emphasizes the choices people may make in selecting resources.

Module 7 addresses the process component of the system and looks at three types of processing: processing energy, materials, and information.

Module 8 introduces concepts of sensing and control. Thus, the feedback loop is studied. Concepts of open- and closed-loop control are introduced.

Module 9 provides an understanding of the impacts of technology on society from a local, national, and global perspective. Students will assess current and future technological systems in terms of their social and environmental impacts.

Module 10 offers students an opportunity to synthesize and apply their knowledge of how the systems model itself can be used as a problem solving tool. Students are encouraged to use systems thinking to design a functioning technological system in response to an identified need.

3.6 Technology Learning Activities

The Technology Education program is taught in a laboratory setting, where students are engaged in hands-on, design-and-construct activity. Active learning characterizes the program. There is very little "chalk-talk."

In fact, one of the most unique facets of the New York program is the evolution of an activity format known as the Technology Learning Activity (TLA). The TLA serves two major functions: First, it insures that the activity is driven by the concepts, skills, and attitudes identified in the syllabus. Second, the TLA's identify a series of "constants", among which are those that directly relate middle school mathematics, science, and social studies concepts to the study of technology. These concepts have been specifically identified by subject matter specialists. The constants serve to provide genuine interdisciplinary connections to other school subjects, the world of work, and to life's experiences.

A distinct emphasis is placed on the interdisciplinary nature of Technology Education as an integrating discipline.

3.7 Senior High School Technology Education

Technology Education is an elective program at the high school level in New York. A full complement of over 30 courses may be taken as stand-alone electives, or can be clustered into 3- or 5-unit sequences for students at all academic levels. Programs may be designed with concentrations in areas of Communications, Construction, Electronics, Manufacturing, Transportation, and Pre-engineering.

Students may fulfill high school Regents or local diploma requirements by taking State-approved 3-unit or 5-unit Technology Education sequences. Certain high school Technology Education courses such as Design and Drawing for Production, Technical Drawing, Architectural Drawing, and Graphic Communications may be used by students to fulfill Art/Music graduation requirements.

One particular high school elective is gaining favor in many schools. The course, Principles of Engineering, is a one-year long course designed to stimulate interest in, and promote equal access to, engineering careers. Additionally, the course introduces students to engineering thinking, a methodology that extends beyond engineering itself, and is transferable to the solution of problems in many different contexts. The course, which addresses several engineering concepts (i.e., design, modeling, optimization, technology/society interactions, and engineering ethics), is taught through a series of engineering case studies. The case studies relate to such topics as ergonomics, auto safety, computer automation, structural design, and energy efficiency. Five engineering colleges in New York are assisting the high school teachers by providing inservice education and networking opportunities. The National Science Foundation has recently (1992-94) funded a New York State Principles of Engineering initiative.

4   Assessment in Technology Education

To fulfill sequence requirements in Technology Education, students must take and pass one 50 question proficiency examination in one of the Systems courses, when available. Presently Systems tests have been developed in three courses (Communications Systems, Production Systems, and Transportation Systems).

There is presently no State examination for the mandated Introduction to Technology middle school program.

The concept of portfolio assessment in gaining favor in Technology Education. Traditionally, these programs (and their Industrial Arts precursors) have always included an assessment component that rated actual student project work. The current trend in portfolio assessment, stimulated by efforts under way in the United Kingdom, includes more of a comprehensive documentation of the thought processes students undertake en route to a problem solution.

5   Issues Impeding Implementation

There are a number of concerns that must be openly confronted before a climate favorable to program implementation can exist. Historical traditions of the teachers, perceptions of the general public, and the lack of clarity of mission impede the implementation of Technology Education programs. These issues must be debated and clarified before positive attitudes can be engendered.

6 Historical Traditions

The traditions of Technology Education spring largely from craft teaching. Most recently, the Science Education community is turning more attention to Technology Education. The National Science Foundation in the United States has identified Technology Education as a funding priority [8]. The American Association for the Advancement of Science's (AAAS) *Science for All Americans* report identifies benchmarks for mathematics, science, and technology for all students from primary school through high school [9]. Since so many teachers have received their undergraduate education through traditional teacher preparation programs, significant inservice education is necessary to philosophically reorient and retrain the bulk of teachers who would teach Technology, as Technology Education is very different from craft-based programs and from science programs in terms of its mission, content base, relationship to other disciplines, and emphasis on design and problem-solving skills [10].

The retraining of teachers is not a trivial task. In addition to providing updated content and methodology, the teacher retraining program must protect the self concept and ego of teachers. Many of these are good teachers, who have been teaching craft-based subjects that were precursors to Technology Education, and teaching them well, for years; and who have, to some measure, defined themselves by their professional affiliations.

For attitudinal change to occur in teachers, the result of transition must be perceived as advantageous. To set the stage, several elements need to be put in place. Among these are a clearly articulated mission, some promise of improved conditions, resources for materials and equipment, and administrative support. If carefully conceived, the change process can leave teachers feeling a new dignity and a heightened sense of purpose.

7 Public Perceptions

Subjects like Mathematics and Science have longstanding disciplinary traditions, and a well established base of support within schools and in industry, government, and the public sector. Technology Education on the other hand, is stigmatized by its origins. Because of its roots in the craft tradition (Sloyd in Sweden, Manual and Industrial Arts in the United States, CDT in the United Kingdom), Technology Education is challenged to substantiate its academic rigor. The perceptions of opinion leaders and policy makers must be shaped by clearly articulated rationales, and overt expressions of support from influential individuals and groups.

In the United States, the general public is yet largely unaware of the existence of Technology Education. Not only does this lack of awareness narrow the base of political and financial support, but mitigates the flow of new teachers into the profession, at a time when many teachers are nearing retirement age, and recruitment of teachers is essential to the survival of the discipline.

8   Current Trends in New York State

The Commissioner of Education in New York, Dr.Thomas Sobol, and the New York State Board of Regents, have developed a comprehensive plan for educational reform in New York State. The plan has been outlined in a document entitled A New Compact for Learning. As a means of implementing this plan, Commissioner Sobol has convened Curriculum and Assessment Committees, charged with developing sets of student Content Standards in the following areas: English Language Arts, Social Studies, Technical and Occupational Studies, Languages other than English, Arts and Humanities, and Health, Home Economics, and Physical Education. One Committee has been formed to generate standards in Mathematics, Science, and Technology Education (MST). The MST Curriculum and Assessment Committee has been meeting for about two years. It has generated a Framework for Curriculum Development that includes nine K-12 Content Standards in Math, Science, and Technology, and also has developed a set of Performance Indicators, that offers more specificity as to what students will be expected to do and to know, at three levels of schooling (elementary, intermediate, and commencement).

The new Framework breaks new ground in education. It opens the door wide for collaborative, interdisciplinary education, where Technology Education plays a central role as an integrating discipline through which mathematics and science might be applied to real-world problem solving. Hands-on, design and construct instructional strategies will be used to convey integrated math, science, and technology concepts and skills.

It is now generally recognized that we are answering the "why study the math, science, technology" question now very differently than 30 years ago. Thirty years ago, each of these were subjects for a small fraction of students. Today, popular consensus is building that these are essential to the fundamental education of all students. Furthermore, the study of mathematics, science, and technology should enable students to do far more than compute numbers or recall scientific facts. It should:

− Add powerful and special ways of asking questions, seeking answers and designing solutions to students' full spectrum of problem-solving strategies.

– Enable students to think in terms of systems and to analyze problems holistically.
– Prepare students to deal with today's information explosion by retrieving, analyzing, and synthesizing information.
– Offer students access to, and understanding and effective use of the products of MST for pleasure and human development, communication, transportation, shelter, security, health, environmental stewardship, and work.
– Help students make connections between products and processes and be introduced to and deal with "Big Ideas".
– Enable students to effectively to use processes and products of mathematics, science, and technology for making informed decisions on personal and public levels.
– Better prepare students for jobs and careers.

Through empowering and preparing citizens in these ways, and by stimulating their interest in continuing MST learning, people's personal needs as well as critical global needs will be served--provided, as specified by the Committee, the Content Standards are supported by assessment.

The MST Committee is concerned with assuring that all children graduate from high school with a basic literacy in mathematics, science, and technology. Its recommendations represent the essential elements necessary for achieving this literacy and for encouraging students with interest and capabilities to pursue careers which advance knowledge in these areas.

To make the necessary changes, we need to clarify our new understanding of what MST has to offer all students within the curriculum. The Committee's establishment of nine overarching Standards relates not only its judgment, but that of many national and international studies and reports.

These Standards are as follows:

STANDARD I: Students will use mathematical analysis, scientific inquiry and engineering design, as appropriate, to pose questions, seek answers, and design solutions.

STANDARD II: The student will acquire an understanding of the basic concepts of systems and their uses in the analysis and interpretation of complex interrelated phenomena in the real world, within the context of Mathematics, Science, and Technology.

STANDARD III: The student will use a full range of information systems including computer systems to handle information; communicate information; model and simulate natural and human made phenomena; measure and control objects, processes, and systems; and understand applications and effects of information within the context of Mathematics, Science, Technology Education.

STANDARD IV: The student will demonstrate knowledge of science's contribution to our understanding of the natural world. To achieve this, the student will

demonstrate understanding and knowledge of scientific concepts, principles, and theories pertaining to:
1) the physical setting,
2) the living environment,
3) the human organism,
and will appreciate the historical development of these ideas.

STANDARD V: The student will acquire the knowledge, skills, and values relative to the tools, materials and processes of technology to create products, services, and environments in the context of human endeavors such as bio-related technologies (i.e., agriculture, health), manufacturing, construction (i.e., shelter and other structures), transportation, and communication.

STANDARD VI: Students will understand and use basic mathematical ideas, including number sense and numeration concepts, operations on numbers, geometry, measurement, probability and statistics, algebra, and trigonometry; be familiar with their uses and application in the real world through problem solving, experimentation, validation, and other activities.

STANDARD VII: Students will identify the interconnectedness of Mathematics, Technology, and Science, and will recognize and use themes that are common to all three disciplines.

STANDARD VIII: Students will apply integrated knowledge of mathematics, science, and technology to solve interdisciplinary problems and make more informed decisions relative to personal and societal concerns.

STANDARD IX: The students will develop habits of the mind, and social and career-related skills in Mathematics, Science, and Technology Education classes that will enable them to work cooperatively with others; achieve success in a post-secondary school setting; enter the workplace ready to adapt successfully to different jobs, and possess skills necessary for continuing advancement.

Integrated Content Standards are emphasized in the goals of the new educational system – integrated because all ways of knowing, and all disciplines – are integrated in the real world. No single subject in mathematics, science, or technology exists or is used in isolation from the others. In practice, these disciplines all reflect ways of dealing with the integrated system – family, community, and the world – in which we live. Therefore, it is important that formalized teaching in mathematics, science, and technology be delivered in the framework of the real world.

Mathematics, science, and technology must be integrated to emphasize their inherent interrelationships and their direct relationship to reality. Through integration of mathematics, science, and technology, students will be helped to see how the various parts of the world are related. The presentation of mathematics, science, and technology as parts of a fundamental base of knowledge, and as methods of thinking and knowing, instead of as discrete disciplines, will help to develop the intellectual concepts and skills students need to understand the underlying connections between seemingly unrelated observations and events in daily-life experiences. For example, rainbows, speech patterns, and musical notes are all vibrations with complex waveforms of different frequencies, and, through mathematical analysis, can be broken down into their basic sine wave building blocks. Social and natural phenonmea can be studied and understood using statistical techniques as well as technical tools.

Learning only becomes meaningful and interesting if students see connections between what they already know and what they are trying to learn. Mathematics, science, and technology, which are complex, and often contain esoteric concepts, need to be taught within contextual frameworks that provide students with the necessary reference points to connect the new learning. Such a context not only provides the linkage with past learning, but also makes the new learning useful for building upon. Within the subjects of mathematics, science, and technology, technology is a natural integrator for mathematics and science and should be used liberally for that purpose.

9   Staff Development in Technology Education

Since the transition from Industrial Arts to Technology Education has created such major change in the discipline, teachers have been challenged to assimilate new procedures, content, and methodology to a greater extent than perhaps ever before in their careers.

Staff development has become a priority for the State Education Department, and for other professional entities (i.e., Teacher Education Institutions, and Professional Associations) concerned with the preparedness of Technology Education teachers.

In conjunction with The New York State Education Department, the Hofstra University School of Engineering has organized a New York State Technology Education Network (NYSTEN). The NYSTEN project, funded for three years by the National Science Foundation, will comprise 120 Mathematics, Science, and Technology Education teacher-leaders in 24 geographic regions throughout New York State. Over the next three years, NYSTEN will provide turnkey inservice training for thousands of Technology teachers in areas of technical content, and pedagogy.

The State Education Department has also supported inservice programs for teachers during the last several years. For example, during the last two years, 50 teachers were trained to deliver the new Principles of Engineering course during a series of 2-day Friday/Saturday intensive meetings. Others have become skilled in the use of computer-aided design software.

There are three Colleges/Universities in New York State that prepare Technology Education teachers. These are Buffalo State College, The City College of

New York, and The State University of New York, College at Oswego. Each of these institutions has offered at minimal cost, summer inservice programs for Technology teachers relating to newly developed State-approved courses, technical content areas, and teaching methodology.

Finally, the Professional Teacher Associations (New York State Technology Education Association) and its Regional Affiliates, sponsor workshops, and Conferences devoted to teacher enhancement.

10 Summary

In New York, a new school discipline is emerging. Technology Education is taking its place alongside longstanding Mathematics and Science programs. Technology Education has the potential to actively engage students in design-and-construct activities that help them to synthesize the knowledge and skills in math, science, social science, and other subjects. As an emerging subject area in the school curriculum, Technology Education faces several issues that impede implementation. These issues relate to its roots in the craft tradition, its relatively recent evolution, and the present lack of public awareness and support. Current trends in New York focus on the use of Technology Education as a vehicle through which integrated Mathematics, Science, and Technology (MST) programs can be implemented.

References

(NATO ASI Series F volumes are indicated by F and volume number)

1. NYSED (1989) Education in New York State: A new basic for the 21st century. Albany, NY
2. Department of Education and Science (1991) The National Curriculum in Design and Technology. Her Majesty's Stationery Office, London
3. NYSED Technology Education Working Group (1991) Outcome-based technology education: A new framework for the study of technology (draft)
4. Hacker, M. (ed.) (1991) Integrating advanced technology into technology education. F78
5. Bowyer, J. (1990) Scientific and technological literacy: Education for change. UNESCO
6. Reading, 'Riting, 'Rithmetic – and now Tech Ed. Business Week, October 18, 1987
7. NYSED (1986) Introduction to Technology. Albany, NY
8. Program Announcement (1991) Directorate for Education and Human Resources. Division of Materials Development, Research, and Informal Science Education. Washington, D.C.: National Science Foundation

9. Rutherford, F.J., Ahlgren, A. (1989) Science for all Americans. Washington D.C.: American Association for the Advancement of Science
10. Hacker, M. (1990) Applying leadership and strategic planning techniques. The Technology Teacher 49(4). Reston, VA: International Technology Education Association

8

Challenges and Lessons Learned in Second Language Learning Systems

Merryanna L. Swartz

Vitro Corporation, 14000 Georgia Ave.
Silver Spring, MD 20906, USA

Abstract: Advances in Intelligent Computer Assisted Language Learning (ICALL) over the past few years have yielded several working systems that promise to do much to change how we learn languages. Some of these efforts have solved problems such as use of multimedia environments that provide learners with robust context and multimodal information to support immersion in a language. Other efforts have shown that problems remain, such as modelling discourse processes and parsing a learner's language for appropriate error feedback. This paper will discuss selected ICALL efforts in the United States and Europe that illustrate the lessons learned and resulting research progress. In addition, several key research challenges in second language learning in advanced technology systems will be presented. This background will provide a state-of-the-art overview of research issues related to Advanced Educational Technology for language learning. In addition to the author's ICALL research, examples will draw from laboratories at Carnegie Mellon University, MIT, Xerox PARC, with mention of user modelling efforts at the Université de Clermont and Université di Udine.

Keywords: Second language learning, CALL, ICALL, natural language processing, multimedia

1 Introduction

Over the past five years, research in advanced educational technology for second language learning has evolved from basic computer-assisted language learning (CALL) systems to more sophisticated intelligent CALL (ICALL) that utilizes artificial intelligence techniques for natural language processing and learner modeling. Some of the progress in ICALL is due in large part to advances in multimedia software tools for designing CALL and ICALL. Multimedia allows direct manipulation of a variety of instructional components that results in the development of enriched, interactive learning environments.

There are several interesting multimedia systems for language learning under development that show promise for second language learning (SLL). However, we shall see that a challenge exists for designing CALL/ICALL with these tools that meets communicative and instructional objectives, and results in effective skill learning. Some of the lessons learned in ICALL research show that development of sensitive natural language processing techniques to capture and correct learners' inaccurate inputs remain the greatest challenge we face.

In this paper, we discuss some of the research issues and progress that has been made as a result of developing advanced educational technologies for SLL. The issues will be organized under the two technology themes of this workshop: Instructional Design and Interactive Systems. The Instructional Design theme presents a current instructional approach for teaching language communication skills. From this background two key technologies will be presented for developing interactive ICALL systems. The research issues involved in both of these areas will be described in order to understand which technologies require further investigation. In the last section of this paper several current CALL and ICALL systems will be reviewed in order to identify some of the lessons learned from research efforts in educational technology for SLL.

2 Instructional Design

Current teaching strategies in the United States focus on the communicative approach to language learning (O'Maggio 1986). Communicative language learning stresses the role of language for communication and transmitting meaning rather than the memorization of grammatical structures. In communicative language learning instructional design, we might say that semantics takes precedence over syntax. This is not to say that syntax is to be ignored, acquisition of grammatical competence is indeed very much a part of SLL and the communicative approach.

The goal of creating a learning system that promotes communicative language competencies should be based on a design that specifies appropriate situations and experiences in which designated language elements occur. The instructional design must always present the language elements in context. The primary assumption behind the design of CALL/ICALL is that language is learned from the environment and that the elements composing the environment in which the learner acquires the language can be identified. This follows from another assumption that SLL is experiential and that these experiences can be arranged to facilitate learning. Thus CALL/ICALL must be designed so that language communication, including extralinguistic components of the language environment, can be practiced and experienced in realistic situations. This communicative context helps learners form a mental representation for the language.

Instructional design for CALL/ICALL, especially a design espousing the communicative approach, should ideally address all four skill areas (reading, writing, listening, speaking). This ordering does not follow the natural progression of language skill learning, but instead the order best supported by CALL/ICALL systems. In order to reinforce those skill areas covered in some CALL/ICALL system, the designer must also include appropriate structural drills. Thus communicative language learning (semantics) does not reject grammar drills (syntax) as part of the curriculum, but instead uses grammar learning to reinforce language patterns encountered in the communicative acts presented in the discourse passages and dialogues used in a lesson.

How then do we create CALL/ICALL so that learners can engage in meaningful language experiences and acquire appropriate, sequenced language patterns that they can use to communicate with others? Discourse narratives are one example. Narrative stories can be designed to define specific discourse patterns for a set of language acts in CALL lessons. The story construct provides situations in which one engages in discourse (the dialogue) for acquiring speaking/listening skills or in a narrative passage for acquiring reading/writing skills. The story context directly affects what linguistic realization is required and how it is to be understood in terms of social roles, topics, events, and pragmatics (Swartz & Russell 1989). These features in theory form the structure for a particular mental representation a learner acquires when learning a language. The CALL/ICALL instructional designer constructs the story contents, the discourse or narrative passage, and the particular language patterns to be learned. This aspect of SLL instructional design is not new. However, specification of the narrative contents are critical in the design process if certain speech acts and linguistic competencies are to be taught. The challenge we now face in ICALL systems is how to represent the language taught in the computer so that learner interaction can be modeled. Story schemas may be one way, grammar representations for parsing language is another.

Deciding what linguistic patterns to teach in CALL/ICALL involves identification of the pedagogical grammar; the subset of the language to be taught. The specification of the pedagogical grammar limits the coverage of the language that needs to be represented in the learning system. Scope of the pedagogical grammar will help the designer construct narrative texts or discourse passages for those speech acts in the grammar, and appropriate structural drills that practice those acts and their linguistic realizations. The scope of the pedagogical grammar will also help the computer programmer develop adequate linguistic coverage by constraining the lexicon and grammar to what is needed in the instructional lessons. We will see later in the paper that this design decision will facilitate the natural language processing required by the system.

3   Interactive Systems

The challenge for implementing the communicative approach to language learning and representing the pedagogical grammar can be addressed by two advanced technologies: multimedia and natural language processing (NLP). The first technology has gained considerable maturity over the past few years. Several authoring tools exist (AuthorWare Professional and Icon Author) that make multimedia CALL[1] design a relatively easy process. Multimedia CALL can support acquisition of all four language skills (listening-speaking-reading-writing). But some of these skill areas such as writing may be limited unless NLP technology is involved. Adequate speaking skills will be limited in CALL until speech recognition technology matures.

Multimedia courseware development tools such as those listed above provide the means to integrate audio, video, and text for presenting realistic discourse passages and narratives, to provide redundant encoding of linguistic material in several cognitive channels, and can illustrate extralinguistic cues (e.g., gestures, eye movements) to enhance acquisition of sociocultural aspects of language heretofore not possible on computers. Hypertext is also very much a part of these multimedia toolkits. The direct manipulation of instructional material in hypertext multimedia systems provides learners with rich, interactive environments that support immediate exposure to language. Some audio tools can even be programmed to provide learners with the means of recording their own spoken inputs to compare with the native speaker. This multimedia feature will allow autocorrection of pronunciation errors in CALL.

Interactive systems in CALL and ICALL must provide learners with appropriate, timely feedback to errors to be effective. Both AuthorWare Professional and Icon Author provide non-AI methods for assessing skill performance and providing feedback. These methods include the means to construct templates for structured pattern drills, multiple choice tests, cloze exercises and the like. More sophisticated interactive techniques that model learner behavior in ICALL are under development, with especially nice techniques emerging from European ICALL researchers (Chanier et al. 1992; Tasso et al. 1992). However, in the current state of the art for ICALL, learner modeling using AI techniques remains a research issue yet to be solved. This issue is recognized as important for developing interactive systems that can provide truly adaptive instruction and sensitive error feedback. However, error modeling will not be addressed further in this paper.

---

[1]   I refer to CALL when discussing multimedia systems that have no other advanced technologies such as artificial intelligence (AI) for learner modelling or NLP for real time parsing. There are of course multimedia ICALL systems (Yazdani's system), but in general the AI/NLP component is the distinguishing technology between CALL and ICALL.

NLP is the second advanced technology needed to develop interactive ICALL that adheres to the communicative approach to language learning. By introducing NLP into ICALL, we can have the learner take an active role in practicing language usage on the computer. In contrast to multimedia technology, however, NLP is not as mature. Its implementation presents a real challenge for ICALL designers. Some of these challenges are how to represent the pedagogical grammar (what theory to use) and how to parse learners' imperfect inputs, their 'interlingua', in such a way so as to provide corrective feedback. Computational linguists develop NLP methods that are closely tied to the implementation of some linguistic theory (Lexical Function Grammar-LFG, Government Binding-GB, etc.). In SLL, selecting the portion of the grammar to be used is done uniquely for instructional purposes and should be tied to the goals of the instruction (e.g., teaching present or future tense forms, subjunctives, etc.). This subset of the language can be called the pedagogical grammar used in some CALL/ICALL system. This instructional objective should make the NLP specialists job somewhat easier by reducing the grammar to be represented. Nevertheless, the labor-intensiveness of providing adequate linguistic coverage in ICALL remains an issue designers must face.

Since computational linguistic theory drives NLP research, different parsing techniques are based on different theories. We can note here that second language research has yet to define a definitive theory for second language acquisition, so for the present, ICALL with NLP uses existing theories such as LFG and GB, for example. At this time, it is unclear how these theories contribute to second language acquisition. Many NLP theorists base their work on first language learning (Kay, 1985) but nevertheless provide useful insights to SLL. Traditionally, researchers from SLL do not interact with computational linguists. Through interdisciplinary interaction, it may be that we will learn more about second language acquisition theory as a result of implementing existing linguistic theories in ICALL. This possibility is an exciting one, and one researchers should attempt to foster in the laboratory.

What is critical in NLP for ICALL is the coverage for learners errors and the ability to provide feedback. SLL errors are unique linguistic patterns that portray predictable errors learners exhibit during the acquisition process. These perturbations can be either semantic or syntactic in content and may render the input unparsable. Yet it is the ability to understand these perturbations and correct them that is of interest. The ICALL system must be able to dynamically identify specific error patterns and provide appropriate feedback in order to ensure learning. This is not an easy problem to solve.

With this brief overview of critical advanced technologies for CALL/ICALL, we next will review current research systems to see what lessons we have learned and to gain insight on where ICALL research needs to focus in the next few years.

4   Lessons Learned in ICALL Research

4.1 Multimedia CALL

There are many multimedia CALL systems that show promise for SLL and the design of the communicative approach to language learning. Three CALL systems that have demonstrated the power of multimedia for designing language instruction include a suite of published videodiscs in MIT's Athena Language Learning Project, an interactive environment developed by the author for advanced vocabulary learning in French (Swartz 1992) and English as a second language, and a dialogue-based CALL system with parser (Pollard & Yazdani 1990). Each CALL system uses multimedia to implement the communicative approach to SLL.

The Athena Project has been an MIT initiative that began in 1984. The published CALL videodiscs (Yale Press) provide interactive narratives and documentaries, both in French and Spanish respectively, for learning the language and culture of the target language. The French language version, *A la rencontre de Phillipe*, presents narratives filmed in France that illustrate regional expressions and feature non-actors in the roles so that speech patterns and phonology are more realistic. Tools allow the learner to interact with the narrative, repeat video segments for reinforcing listening skills, and practice repeating the dialogues. Online exercises provide grammar correction for sentences created by learners in a limited domain of pre-structured dialogues.

The ability to see interlocutors interact in realistic video scenes presents extra linguistic cues and sociocultural features of language that heretofore were difficult to present with high fidelity on the computer. The power of multimedia to convey this information instantaneously, and to allow 'instant replay' of selected video segments makes multimedia CALL design a must for every classroom and language laboratory.

LEXNET is another multimedia environment designed to teach French vocabulary to advanced learners. LEXNET was developed as an instructional approach to vocabulary learning based on a psychological theory of how lexical information is stored in memory (Miller & Fellbaum 1992). This CALL presents vocabulary in hypertext semantic networks that provide a context for the words along different semantic relations (synonymy, antonymy, meronymy). These relations follow Miller's work to develop an electronic dictionary with linkages for various semantic relations to other words in the dictionary. In LEXNET, the words are also presented in narrative text so that learners can see how networked words are used in a natural language (communicative) context. These two different contexts for words are designed to support the acquisition of enriched mental representations for lexical information in a second language. Sound is included so that learners can hear how words are pronounced. Figure 1 illustrates LEXNET's multimedia interface for an English version. An experiment to compare the network context with

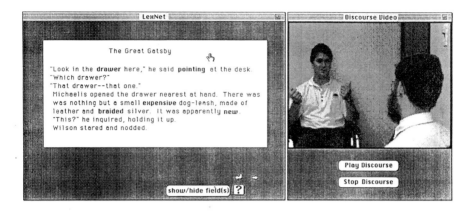

Figure 1. The video window presents the aural dialogue shown in the narrative text window. Bolded words are members of the lexical network structure. A pull-down menu lets learners see a translation, a grammar note for the passage context, word pronunciation, and a tool to take learners to the network structure for that word.

the narrative context indicated that semantic networks improved learner's ability to recall words from memory (Swartz 1992). Thus, while the communicative approach reinforces the use of language to communicate, vocabulary learning, especially for advanced learners, may benefit from network structures for words. In the English version of LEXNET currently under development, video is included to provide live depiction of the narrative or dialogue.

LEXNET uses hypertext, online word activities, and cloze exercises to provide an interactive CALL system. Miller's online dictionary (Miller 1992) is being integrated as a word lookup tool to support the network presentation of vocabulary and help learners expand their lexicons in the target language.

Although *A la rencontre de Phillipe* and LEXNET use no AI techniques, the richness of multimedia coupled with appropriately designed error correction and feedback make these environments good examples of what is possible with today's advanced technology. The third multimedia system we discuss adds NLP to illustrate how a parsing algorithm can enhance system interaction during instructional exercises.

A NLP-based multimedia CALL developed at the University of Exeter, eL (Pollard & Yazdani 1990), demonstrates the next level of interactive CALL that permits learners to practice natural language generation in the target language. eL meets the definition of an ICALL system, but it is presented here since it combines multimedia with NLP to bridge the gap between CALL and ICALL systems. The eL system uses sound and graphics to present instructional dialogues in context in either French, Spanish, or English. Graphical, animated scenes present the communicative components that surround the dialogue with tools that

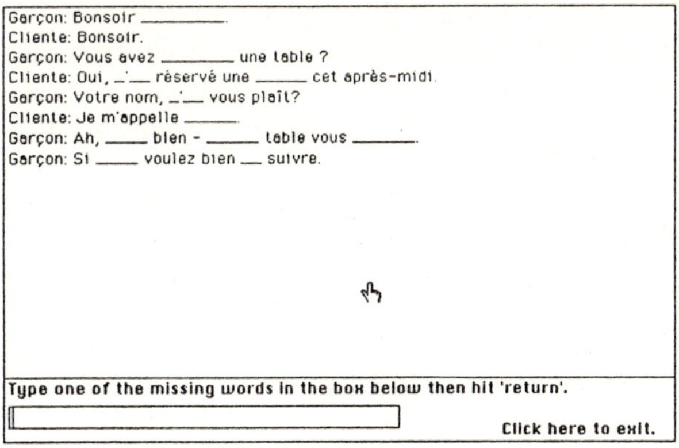
Garçon: Bonsoir ______.
Cliente: Bonsoir.
Garçon: Vous avez ______ une table ?
Cliente: Oui, _'__ réservé une ______ cet après-midi.
Garçon: Votre nom, _'__ vous plaît?
Cliente: Je m'appelle ______.
Garçon: Ah, _____ bien - ______ table vous ______.
Garçon: Si _____ voulez bien ___ suivre.
Type one of the missing words in the box below then hit 'return'.
Click here to exit.

a.

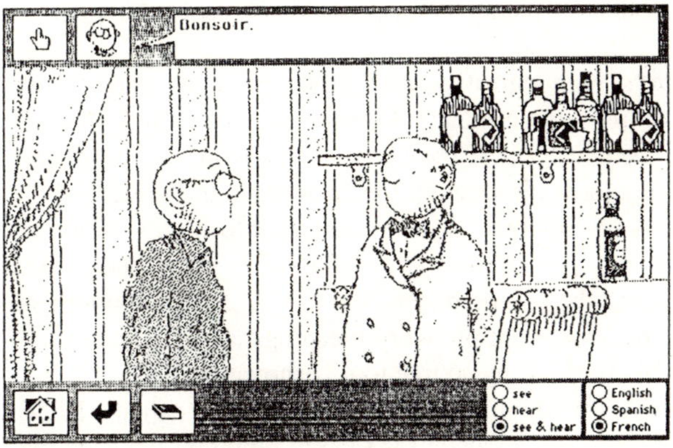
Bonsoir.
see
hear
see & hear
English
Spanish
French

b.

Figure 2. a the graphical scene for a restaurant dialogue. Tools let learners see and hear dialogue components. **b** a cloze exercise. The PROLOG parser analyzes each sentence when it is completed and presents simple error feedback.

let learners hear discourse passages as they view the environment in which the communicative acts occur (see Fig. 2a). Online exercises shown in Fig. 2b illustrate traditional cloze activities for the instructional dialogue. The innovation in eL is the use of a PROLOG-based parser to check the grammar of completed sentences. It differs from the ICALL systems we discuss next in that the eL parser follows no formal linguistic theory.

4.2 ICALL

ICALL systems are less prevalent than multimedia CALL, however, there are many active research groups in both the United States (Crosby & Stelovsky

1993[1]; Evans & Levin 1993; Rypa & Feuerman 1993) and Europe (Chanier et al 1992; Tasso et al. 1992; Yazdani 1989) that are actively engaged in NLP research in SLL. The two exemplar ICALL systems we discuss next have demonstrated the success of integrating NLP in instructional language learning systems. Each is briefly presented below.

CALLE (Rypa & Feuerman 1993) is an ICALL system in Spanish that was designed to teach reading and text comprehension skills to advanced adult learners. It is a language analysis tool with NLP technology that lets learners explore various aspects of narrative text. The linguistic theory CALLE uses is LFG, a linguistic representation and parsing algorithm that is independent of any one language but focuses on the representation of language functions. As a theory, LFG analyzes two structures for a parsed sentence, a constituent structure that defines the surface form for some phrase, and a functional structure that specifies the grammatical functions for each word (e.g., subject, object, etc.). These two representations and their graphical depictions may provide instructional support to adult learners.

CALLE presents learners with narrative text passages in an easy-to-use graphical user interface with pull down menus. Learners have a variety of tools they can access to examine the text further. They can select a word and get access to various lexical and morphological information for the word in the context of the text passage. A sentence look up tool provides explanatory information based on the parse of the sentence. A third tool shows a modified LFG parse tree for the phrase selected. These tools and the exploratory nature of CALLE let learner find out more about a particular passage, multiple word meanings, and the nuances of language in context. A sample CALLE parse tree tool is shown in Fig. 3.

Another NLP-based ICALL for learning elementary Japanese is called ALICE-chan (Evans & Levin 1993). This system also uses LFG to represent the pedagogical grammar, but unlike CALLE, results from the parse are not made available to learners. In ALICE-chan, the instructional approach focuses on presentation of basic dialogues with translation and cloze exercises. Lessons are presented in kanji and learner responses are accepted in kanji or romanji. Figure 4 shows a typical screen from the ALICE-chan interface.

Both ALICE-chan and CALLE have been used in schools or language centers with actual students. Each system has shown promise for providing useful language instruction. Although these systems support reading and writing skills, they do promote the communicative approach through presentation of language in narrative texts and discourse passages. These ICALL systems have concentrated on the integration and use of NLP to parse learners inputs (syntax) within the context of instructional dialogues or defined narrative texts (semantics). In this regard, both systems appear to counter the instructional focus of the communicative approach

---

2   This Japanese ICALL system also uses multimedia to provide an enriched, interactive learning environment.

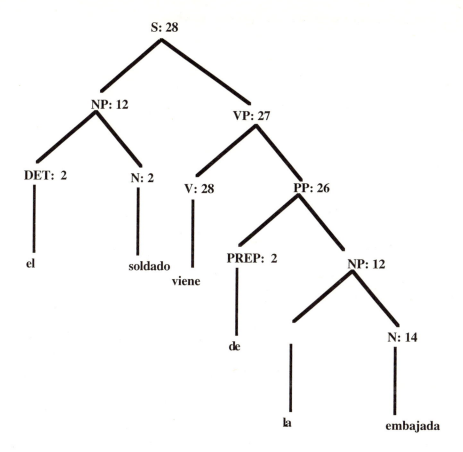

Figure 3. A parse of a simple Spanish sentence encountered in a narrative text passage

by supporting syntax over semantics. However, we have stated earlier that acquisition of grammatical competencies is a key component of effective communication. What remains a research issue is the assessment of how and when to use grammar-focused ICALL in the curriculum.

While ALICE-chan provides limited error feedback, it does not yet include a robust user modeler of the type typical in intelligent tutoring systems. Independent SLL research groups have been investigating this modeling technology for use in ICALL. Some systems like the French tutor under development at the University of Clermont (Chanier, et al., 1992) are developing ways to model learners ill-formed inputs. Another effort in an Italian laboratory (Tasso, Fum & Giangrandi, 1992) is developing modeling algorithms for analyzing learners' interlingua. Both of these systems have NLP technology, but

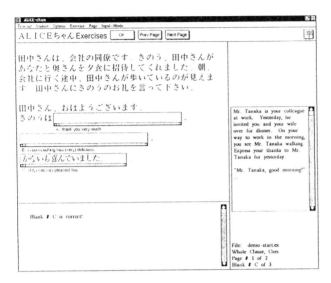

Figure 4: A multiwindow presentation from a translation exercise

focus more on the learner modeling aspect of interactive systems than on language instruction. This research can help us better understand what the meaning and patterns of learners errors reflect about the progression of skill acquisition in the target language. User modeling research shows promise for developing more robust theories of SLL.

5   Conclusions

These brief overviews of multimedia CALL and NLP-based ICALL show that progress over the past five years has been considerable. Multimedia tools for authoring computer-based instruction are now readily available. Second language teachers can use them to author courseware as evidenced by the recently published programs completed through the Athena Project.

Engineering CALL with the authoring tools that exist is one of the lessons learned in this field. The critical issue that remains is to understand how to design the instruction so that communicative competencies can be taught and practiced. Another research issue that is critical is evaluation of instructional effectiveness. The simple availability of multimedia tools will not ensure valid instruction if we do not understand how learners learn and which skills to teach when. A research challenge we can easily meet is the formation of multidisciplinary groups so that language teachers work with computational linguistics to design learning environments with sound SLL instruction and appropriate error feedback and remediation exercises.

Challenges for ICALL, despite the recent progress, are ever present. NLP techniques have demonstrated that reasonable linguistic coverage is possible for a defined course and skill level (Levin & Evans 1993). What we need to do is develop better error detection methods. We need to begin integrating parsing algorithms for SLL with error and learner modelling techniques. This is no easy task, nor is it cost efficient at this time. We can draw from some of the techniques developed in intelligent tutoring systems research, but the problem of combining research foci into one ICALL still remains.

It seems plausible to state that in another five years the challenge of integrating user modelling with SLL parsing algorithms will become yet another lesson learned. That is the hope we have. What continues, and was not addressed in this paper, is our fundamental understanding of the second language acquisition process. One recent study that has implications for how we teach languages shows that spaced retrieval practice influences long-term language maintenance (Bahrick et al. 1993). While we are becoming better CALL/ICALL engineers, we still do not know how we acquire second languages. How native languages influence SLL, how we store and access second language information, and the psycholinguistic aspects of bi- and multilingualism are issues we need to investigate further. These are the really big challenges that loom on the horizon of advanced educational technology for language learning.

References

(NATO ASI Series F volumes are indicated by F and volume number)

Bahrick, H.P., Bahrick, L.E., Bahrick, A S., Bahrick, P.E. (1993) Maintenance of foreign language vocabulary and the spacing effect. Psychological Science 4, 316-321

Chanier, T., Pengelly, M., Twidale, M., Self, J. (1992) Conceptual modelling in error analysis in computer-assisted language learning. In F80

Crosby, M., Stelovsky, J. (1993) Hypermedia as a facilitator for retention: A case study using Kanji City. Paper presented at the Reactive and Creative CALL Conference, University of Exeter, UK, September

Kay, M. (1985) Parsing in functional unification grammar. In: D.R. Dowty, L. Karttunen, A.M. Zwicky (eds.) Natural language processing: Psychological, computational, and theoretical perspectives. Cambridge, UK: Cambridge University Press

Levin, L. S., Evans, D.A. (1993) ALICE-chan: A Japanese tutor. Paper presented at the Workshop on Advanced Technology for Language Learning, Alexandria, VA: US Army Research Institute

Miller, G. A., Fellbaum, C. (1992) WordNet and the organization of lexical memory. In F80

Murray, J., Felshin, S. (1993) Advanced technologies and language learning: Lessons from the Athena language learning project. Paper presented at the Workshop on Advanced Technology for Language Learning, Alexandria, VA: US Army Research Institute

O'Maggio, A.C. (1986) Teaching language in context. Boston, MA: Heinle & Heinle

Pollard, D., Yazdani, M. (1990) A multimedia restaurant scenario. Department of Computer Science, University of Exeter, UK

Rypa, M., Feuerman, K. (1993) CALLE: An exploratory environment for foreign language learning. Paper presented at the Workshop on Advanced Technology for Language Learning, Alexandria, VA: US Army Research Institute

Swartz, M.L., Russell, D.M. (1989) FL-IDE: hypertext for structuring a conceptual design for computer-assisted language learning. Instructional Science 18, 5-26

Swartz, M.L. (1992) Issues for tutoring knowledge in foreign language tutoring systems. In F80

Tasso, C., Fum, D., Giangrandi, P. (1992) The use of explanation-based learning for modeling student behavior in foreign language tutoring. In F80

Yazdani, M. (1989) Artificial intelligence techniques in language learning. Proc. 2nd International Conference on Computer Assisted Learning, Dallas, TX, pp. 618-624. Springer-Verlag

9

Student Modelling: A Crucible for Research

Gordon I. McCalla

ARIES Laboratory, Department of Computational Science
University of Saskatchewan, Saskatoon, Saskatchewan S7N 0W0, Canada
 E-mail: mccalla@cs.usask.ca

Abstract: Student modelling is a centrally important issue in the construction of intelligent learning environments and intelligent tutoring systems. Without a student model, it is impossible for such a learning environment to adapt to the needs of individual learners. Unfortunately, student modelling is also a very difficult problem. It touches on many of the great issues of artificial intelligence and cognitive science: diagnosis, belief revision and truth maintenance, qualitative reasoning, mental modelling, temporal reasoning, non-monotonic and probabilistic reasoning, testing and evaluation, etc. Student modelling provides both a focus for the exploration of these issues, as well as an original twist on many of them. The original twist arises due to two main factors that are central to student modelling but are often not important in other applications. The first of these is the impossibility of keeping a completely accurate model of the learner, which forces the student model to deal with inherent uncertainty and incompleteness. The second factor is the constant revision the learner undergoes in his or her perceptions of the domain of study as the instructional interaction proceeds, a feature that presents a constantly moving target for the student modelling subsystem. In this paper I discuss the importance of student modelling, the key issues that must be tackled, and the prospects for resolving these issues in the short and long term.

Keywords: Student modelling, granularity, student model maintenance, individualization

1 Introduction

The more that is known about a learner, the easier it is to tailor a learning environment to meet the individual needs of the learner. With no knowledge of the learner, the environment cannot be individualized at all. Thus, the problem of learner or student modelling is a central one for anybody who wishes to build intelligent learning environments (ILE's) or intelligent tutoring systems (ITS's) that are flexible and adaptable to learner needs. Unfortunately, student modelling is an extremely difficult problem The problem is so hard, in fact, that many in the ITS research community have fled in the face of it. They have transformed the goal of building individualized adaptive tutors centered around a student model to the new goal of building learning environments that provide a range of tools that provide scaffolding for a learner but do not try to track the learner's understanding as these tools are used.

I believe that it is premature to give up on student modelling. There are three reasons. First, there has been progress in student modelling to the extent that it is now possible to provide a significant and practical student modelling component to an ITS. Second, student modelling is unavoidable. Current attempts to build environments without student modelling will sooner or later founder on the need for individualization, much as Logo-style discovery environments need to be augmented by human tutorial assistance if learners are to avoid cul de sacs and are to be challenged into new directions. Finally, it is important to continue to investigate student modelling because to do so is to continue to investigate critical artificial intelligence (AI) and cognitive science questions. Student modelling challenges traditional approaches to diagnosis, belief revision and truth maintenance, qualitative reasoning, mental modelling, temporal reasoning, non-monotonic and probabilistic reasoning, testing and evaluation, etc. I have discussed elsewhere the first two reasons that student modelling is important. The necessity for doing student modelling is argued in (McCalla, 1992); a number of interesting techniques and current student modelling research directions are collected in (Greer and McCalla, 1994). In this chapter I would like to concentrate on this third facet of student modelling: its role in stimulating research in AI and cognitive science.

Student modelling gives many AI and cognitive science issues a focus, thereby constraining these issues somewhat, but the focus also usually provides a new and interesting twist on the issues. Diagnosis of a device in AI becomes cognitive diagnosis of a changing student in ITS, and this is quite a different problem from device diagnosis because there is no a priori correct model of the student (see Self 1992, for a detailed discussion of this issue). Belief revision and truth maintenance take on ominous, but fascinating, complexities in ITS since there are many kinds of inconsistency in a learner and these are changing all the time. Learners are always reasoning qualitatively, they have many layers of reasoning, and their reasoning changes over time. To fully capture the reasoning of a learner therefore requires providing qualitative reasoning formalisms that have cognitive validity; providing a range of mental modelling, analogical, and other reasoning capabilities that can work in synch with one another; and providing a temporal reasoning capacity that not only tracks real events but also more subtly monitors the kinds of hypothetical reasoning and backtracking in which learners engage. In a student model, non-monotonic and probabilistic reasoning must work on incomplete and contradictory knowledge. Testing and evaluation of any learning environment is extremely difficult since no double blind or even single blind experiments can be carried out, and the reasons for success or failure are inevitably obscure.

In order to make any progress on these difficult issues, it is critical that researchers give up several holy grails. The search for completeness must be abandoned: any student model is inherently incomplete. The requirement that a knowledge representation formalism be consistent must be lifted: it is impossible to model a learner with an internally consistent knowledge representation scheme, because the learner is inconsistent. The size of the student model and the fact that it must constantly change to track changes in the learner will also preclude any realistic chance of maintaining logical consistency. The quest for certainty must be abandoned. It is a practical impossibility for a student modelling system to be absolutely certain of a learner's state of knowledge at any given time; at best a rough estimate will be all that is possible. Finally, it no longer suffices to study issues "in the small". Mechanisms to do student modelling must scale up to large knowledge bases and must gracefully degrade as they fail.

In short, when investigating student modelling, a researcher is forced to be realistic, to look at problems in the scruffy real world, to deal with knowledge as it is, not as a logician would like it to be. This means, to use Brown's (1990) terminology that student modelling has better "forcing functions" for the knowledge representation enterprise than do expert systems, natural language understanding, medical diagnosis, or many of the other domains AI folks are fond of. Student modelling is thus a crucible for interesting research.

In this chapter I would like to explore in somewhat more detail how student modelling can stimulate research into interesting knowledge representation questions. I do this by discussing two main research issues: granularity in reasoning and student model maintenance. For each issue, I not only discuss why the issue is important for student modelling, but I also show that it raises deep and interesting questions for AI more generally. The philosophical discussion of each issue is also imbued with some depth by presenting research projects (carried out under the auspices of the ARIES Laboratory) which illustrate the issue. Granularity is illustrated by the SCENT project to provide advice to LISP programmers on their solutions to programming problems. Student model maintenance is illustrated through work carried out by Xueming Huang, a former Ph.D. student in the Laboratory, on the construction of a three part belief revision system for aspects of student modelling. I then conclude the paper with a look at what is practical now, what is coming down the road, and what are the key issues that must still be tackled. First, however, I would like to describe typical functions for a student model, so as to give some background for what follows.

2 What is a Student Model?

A student model is that part of an intelligent tutoring system (or intelligent learning environment) charged with the responsibility of keeping track of individual learners. The idea of student modelling goes a long way back. The earliest computer assisted instruction (CAI) systems would quiz learners on each module (frame) of a course, and depending on their score on the quiz, would choose a new module for the learner to work on. This was essentially a form of distributed student model, based on levels of expertise about specific parts of the domain knowledge. When these CAI systems were made more flexible by giving the learner more control over his or her learning experience, it became important to be able to infer a learner's shifting knowledge states through ongoing background monitoring of behaviour, rather than by explicitly testing them at every turn. This led to a need to be able to represent the domain knowledge and reasoning strategies that a learner was acquiring, to be able to encode a learner's misconceptions, to be able to annotate how well the domain knowledge was learned and how successful was the reasoning, and to be able to diagnose all of these things indirectly as the learner solved problems or engaged in other learning activities. It also became important to discover the goals, motivations, and other non-content aspects of a learner's world view, in order to be able to factor these aspects into a decision about what learning activities to make available to the learner. Finally, it became desirable to keep a history of a learner's progress, or lack thereof, as he or she used the learning environment in order to be able to measure the success or failure of the learning opportunities provided and to be able to make pedagogical decisions based on what seemed to have worked (or failed to work) in the past.

Together, these factors constitute the desirable features of a student modelling component for a tutoring system that is meant to be intelligently responsive to learner needs. In fact, the success with which the student modelling component is able to model the learner is some measure of the intelligence of a tutoring system or learning environment. If the student modelling subsystem is weak or non-existent, then no appropriate individualization can occur. To this extent, then, student modelling is a core problem for ITS.

Unfortunately, to model a learner's knowledge, to infer his or her motivations and goals, and to keep track of the learner's progress are extremely difficult problems, not even approximately solved as yet. Most student modelling systems (see Greer and McCalla, 1994) only try to understand what knowledge the learner currently knows (often called, in the AI sense, the learner's "beliefs"), and have left motivations and history as yet to be solved problems. Even the problem of understanding the current knowledge state of a learner is a difficult one, as will be seen in the following sections of this paper. In Section 3, next, I show how an intelligent tutoring system can make use of granularity to help it to understand the current level of learner understanding, even when it fails to be able to completely diagnose learner behaviour to all levels of specificity. The use of granularity sheds some light on many representation and diagnosis aspects of student modelling, and beyond that illuminates some basic AI problems. In Section 4, I discuss the problem of how new knowledge of learner behaviour can affect an intelligent tutoring system's existing beliefs about the learner and how this leads to a difficult problem of correcting side-effects of this new knowledge, the so-called student model maintenance problem. The student model maintenance problem is a particularly interesting variant of the standard AI truth maintenance problem, and as with granularity above, illustrates AI issues in a particularly interesting light.

3   Granularity

Learners can exhibit both deep and shallow knowledge about the same concept at the same time. For instance, empirical studies in the ARIES Laboratory (Bhuiyan et al. 1989, 1991) have shown that, when learning how to program, learners often have a mental model that underpins their strategy to solve a programming task. This mental model provides a learner with a deep level of conceptual understanding of the problem solving strategy, but this deep level must be translated into more surface forms, eventually resulting in actual code, in order to devise the final solution to the programming task. The range of understanding the learner has, from deep mental model through more shallow levels to surface code, encapsulates several simultaneously held perspectives on the problem solution at three different grain sizes. For example, a learner may visualize that his or her solution to a programming task is a collection of cases at the mental model level, a particular arrangement of loops and tests at the strategy level, and a specific sequence of code in a particular programming language at the surface level. The student model must represent all of these perspectives if it is to reflect the learner's understanding; in short it must be able to handle granularity in representation.

Learners also have only incomplete knowledge of the domain they are learning. Their domain knowledge can range from total ignorance through partial knowledge to fully formed misconceptions. Partial knowledge can come in two versions Partial knowledge may take the form of only knowing about some of the components that make up a concept (e.g., knowing that a platypus has a beak but not that it has hair). Partial knowledge can also take the form of ignorance about the complete details of a concept or procedure (e.g., knowing that there are mammals but not about any specific kind of mammal). In this latter kind of partial knowledge a learner can be said to only have a coarse-grained perspective on the concept. As a learner refines his or her understanding, the learner can, in Hobbs' (1985) terms, articulate his or her knowledge to finer grain sizes. Such articulation seems to happen along at least three dimensions: aggregation (where the learner learns the components of some concept, e.g., that the platypus has hair), abstraction (where the learner learns specializations of some concept, e.g. that the platypus is a mammal), and goals (where the learner elaborates his/her goals into subgoals, e.g., from learning about the animal kingdom to learning about mammals). If the learner moves in the opposite direction, he or she can be said to be simplifying his/her knowledge, from fine grained knowledge of particular situations to an understanding of inclusive, generic, coarse-grained knowledge. In either case, it is important to be able to represent knowledge simultaneously at many grain sizes in order to be able to track a learner's movements whether articulating or simplifying.

Work by Jim Greer and myself in the ARIES Laboratory is aimed at formalizing a notion of granularity and making explicit use of it in a student modelling context in the SCENT program advising system (McCalla et al. 1992). The ultimate goal of the SCENT system is to guide a learner towards understanding the programming language LISP. The idea is to represent possible learner programming strategies in granularity hierarchies, organized along two orthogonal dimensions: aggregation and abstraction. Any particular learner will actually use only a subset of these strategies, and it is up to the tutoring system to recognize which ones, i.e., to diagnose the learner's behaviour. To date, we have not been concerned about representing granularity in a learner's goals.

If a learner is allowed to exercise a good deal of freedom, a system to diagnose learner behaviour will have a difficult time tracking what the learner is doing. In fact, most diagnostic systems are very brittle in the face of unexpected learner behaviour. One approach to dealing with this is to restrict the learner to using one of the system's prespecified models of behaviour. Such "model tracing" tutors, however, tend to force the learner along only those pre-specified paths that the system knows about (as in the early LISP Tutor (Anderson and Reiser, 1985)). If instead, the range of possible learner behaviour and learner knowledge can be formulated in granularity hierarchies, then, while it might be impossible to recognize what a learner is doing at a fine grain size, it usually is possible to recognize what he or she is doing, at least at a coarse grain size. For example, a learner's solution to a programming problem might be unrecognizable as a tail-end recursion (due to extra parts and/or missing parts), but it may at least be recognizable as a recursion. Thus, granularity-based recognition is robust in that learner behaviour is always recognizable at some level of detail. Such robustness is critical to any system if it is to fail soft when it moves beyond its expertise.

Of course, a clear distinction must be made between this kind of coarse grained recognition of a learner's knowledge, which signifies that the system has failed to completely understand the learner, and recognition of a learner's coarse grained knowledge, which means that the learner has not completely understood the domain. Using granularity to represent the incompleteness of learner knowledge is also useful, since a learner can often slowly refine his/her understanding of the domain, in the early stages having only coarse grained understanding of a concept, and only later fleshing out his or her knowledge of that concept.

Another important consequence of using granularity hierarchies for recognition is the fact that partial recognition may occur at a fine grain size at the same time full recognition occurs at coarser grain sizes. For example, the perturbed tail-end recursion above may almost be valid, except for an inappropriate base test. Such partial recognition is important as well, since it can suggest places where the learner may be having trouble, useful information for making pedagogical decision about how (or even if) to help the learner. The two kinds of incomplete knowledge which granularity hierarchies allow, coarse grained full recognition and fine grained partial recognition, thus allow for two different kinds of uncertainty to be represented in the student model, each of which is described qualitatively. This qualitative approach to uncertainty, while less formal than uncertainty calculi like Bayesian statistics, may prove to be much more useful for reasoning than is some number measuring the quantitative uncertainty of a concept. The structures which are partially recognized indicate what is known with certainty, but also show where the certainty breaks down, i.e. the recognized and unrecognized parts are clearly distinguishable, and thus can be reasoned about, perhaps forming the basis for further interaction with the learner. Thus, in some sense qualitative uncertainty not only shows *that* there is uncertainty, but also *what* is uncertain.

Granularity hierarchies are at the basis of SCENT's recognition system which has been implemented and will recognize over 200 common recursive strategies (at various grain sizes, both well formed and misconceived). The SCENT advisor has been beta-tested on real learners, both on a standalone basis and as the back end to Bhuiyan's PETAL system (Bhuiyan et al., 1991), and has worked with robustness and efficiency, although its impact on the learners' problem solving abilities was not tested in anything but an anecdotal way. Granularity thus provides a potentially useful approach for student modelling that is practical right now. Further investigations into the nature of granularity should enhance the usefulness of granularity-based recognition.

In fact, it seems clear that granularity is a central notion in much of AI, not just in ITS, and thus lessons learned in an ITS research project should generalize beyond the field. Granularity-based reasoning would seem to be useful in expert systems, device diagnosis, and in knowledge representation for domains where complete knowledge is unavailable or impossible to enumerate. Of course, many of the granularity notions we have been working with have been originally sparked by research in other areas of AI such as computational vision (Mackworth and Havens, 1981) and formal knowledge representation (Hobbs, 1985). However, ITS provides a great testbed for investigations into granularity since it is so important in the area, and since it seems impossible in principle to be able to fully anticipate the wide range of possible learner behaviours in even the most trivial domain. Thus, it has been through work in student modelling that granularity has come to its fullest fruition. This illustrates the main point of this paper: that student modelling is a crucible for research, not only in ITS but also in AI more generally.

4 Maintaining the Student Model

Another issue I would like to consider is the problem of maintaining a student model as the learner's knowledge changes over time and the tutor's understanding of the learner evolves. It would seem reasonable to accomplish student model maintenance by using a standard truth maintenance system such as that of Doyle (1979) or deKleer (1986) or others. Such truth maintenance systems are able to determine how new information conflicts with existing information in some logic-based model of the world. This is achieved by keeping track of dependencies among facts in the model. When new information contradicts one or more existing facts, the truth maintenance system not only finds the direct conflicts but can also figure out the indirect conflicts by using the dependency relations. The truth maintenance system, however, does not try to derive all the internal implications of its model of the world, so the model, internally, may not be fully logically consistent. This is no problem in student modelling since learners often continue to function with unnoticed internal contradictions. Thus, standard truth maintenance systems at least have some possibility of being used for student model maintenance.

It is important, however, to draw a distinction between beliefs the tutor has from beliefs the tutor has about the learner, i.e. the distinction between the proposition *Student-Believes (Foo)* and *Foo* itself. It makes sense to maintain the logical consistency of propositions that are "outside the brackets" like *Student-Believes (Foo)*, since these are beliefs the tutor has, but it doesn't make sense to maintain logical consistency in the set of beliefs "inside the brackets" like *Foo* since these are the possibly inconsistent beliefs of the learner. That is, it is perfectly reasonable for a learner to both believe *Foo* and not believe *Foo*, but is unreasonable for the tutor to both believe that the learner knows *Foo* and not to believe that the learner knows *Foo*. Thus, truth maintenance techniques can be more easily extended to handle propositions outside the brackets, where logical consistency is plausible, than inside, where cognitive fidelity is the key requirement. I would first like to consider some of the issues involved with maintaining the beliefs inside the brackets, which is the "real" student modelling problem, then move outside the brackets to discuss some work in the ARIES Laboratory carried out by Xueming Huang on the relatively easier, but still useful, task of maintaining the tutor's own beliefs.

When looking at the knowledge of learners, it is important to be able to distinguish between what knowledge they actually have and use, and what knowledge they *know* they have. Learners often have knowledge they don't know they have, although their actions are clearly affected by this implicit knowledge. How can such implicit knowledge be handled by a standard truth maintenance system? If an implicit fact were added to the student model, then it would be able to accurately reflect learner behaviour, although it would not accurately reflect what a learner would claim to know. If, on the other hand, an implicit fact were left out of the student model, then the student model would accurately reflect what the learner would claim to know, but would not be able to accurately reflect learner behaviour. Student modelling thus challenges traditional notions of a knowledge base being logically omniscient, i.e. knowing what it knows, and requires that some mechanism to handle implicit knowledge be devised.

A truth maintenance system does not remove conflicts: it just points them out. A separate reasoning system must choose how to resolve these conflicts. A standard assumption in truth maintenance is that new knowledge is more accurate than old, which means removing old information that conflicts with new information. Such an assumption may not be appropriate in student modelling, given how difficult it is to accurately assess the learner. An apparent change in the learner's behaviour could well be a mistake in diagnosis, rather than a change in the learner. To handle such uncertainty in diagnosis, Elsom-Cook (1988) proposes the idea of bounded student models, similar to the notion of granularity described in Section 3. In Elsom-Cook's approach, the student modelling system can indicate a range in which learner behaviour falls, rather than having to specifically diagnose the behaviour.

While granularity and bounded student models may well begin to solve the problem of uncertainty in diagnosis, their use in student modelling would require that the student model contain non-propositional knowledge. Standard truth maintenance systems work only on propositional logic, and it is unclear how to extend them to work with non-propositional knowledge structures such as we and Elsom-Cook propose. It isn't even clear what constitutes a conflict, since no longer is logical contradiction the hallmark. Pseudo-conflicts arise, where, for example, no conflict occurs between certain concepts at coarse grain sizes (e.g. birds fly but mammals don't), but conflicts do occur among their finer grained descendants (e.g. Sandy the ostrich doesn't fly but Belfry the bat does). Moreover, since much of the information that must be kept in a student model is procedural (e.g. how to solve problems, learner goals and motivations, etc.), the very concept of conflict is open to doubt. Does it make sense to talk about conflict between two procedures? How can two learner goals be said to be in conflict? Like speech acts in linguistics, knowledge of goals and procedures seems inherently non-propositional. There are likely fundamentally different truth maintenance processes from the standard ones needed to handle the kinds of non-propositional knowledge that infest a student model.

The whole notion of truth and falsity, so clear at the logical level in traditional truth maintenance, becomes problematical in student modelling. Learners can often exhibit misconceptions that are not strictly speaking either true or false, just inappropriate. A learner who solves the programming task of averaging a set of 1000 numbers by creating 1000 differently named variables to represent these numbers is not wrong, just terribly misguided. In fact learners can purposefully be given information at a coarse grain size or told incorrect information in order to help them understand an issue (this is Gutwin's idea of a "pedagogically motivated misrepresentation" (PMM, see Gutwin and McCalla, 1992). The information in these PMMs is not true in our world, but the student model must represent them as true for the learner, at least as long at the PMM remains in effect. What happens when the PMM is removed? Thus, student modelling challenges standard notions of truth and falsity.

Goldstein (1982) suggests that any learner's knowledge evolves through genetic transformations. A principled computational theory of instruction based on a deep understanding of the genetic paths through learner knowledge would determine what kinds of instructional actions would be appropriate at a given stage of a learner's knowledge evolution. It would also predict expected effects on the student model of these actions, and would define what kinds of actual learner behaviour would be in conflict with these expected effects. The only way such a principled theory could be devised is by empirical testing of actual learner behaviour, and it may turn out that there are many different types of learners each with their own genetic propensities. What this means for student model maintenance is that there may be many different kinds of maintenance processes, depending on peculiarities of learners. Once again, then, student modelling forces new and interesting research issues to be investigated, in this case the issue of how to model individualized learning styles.

An even tougher maintenance problem arises, however, if the learner's knowledge isn't accumulated in an evolutionary way (predictably through expected genetic transformations in his/her knowledge), but instead progresses in big jumps. Such learners can have long term difficulty in understanding a tough concept, but then can suddenly make a revolutionary change in their perception that radically improves their understanding of the concept. Presumably, the learner has been accumulating exceptions to his/her standard perspectives over an extended period, but has not been willing and/or able to fundamentally alter those perspectives. When the accreted exceptions become too cumbersome, then a radical shift in perspectives can occur, stimulated perhaps by some sort of eureka-like insight. Once the insight occurs, a major reshuffling of the student model must take place, which promises to be much harder than the kinds of incremental change usually associated with truth maintenance.

Another important requirement for truth maintenance in student modelling that may not be necessary for truth maintenance in expert systems, is how to deal with cognitive limitations, such as forgetting. In some ways truth maintenance systems are designed to predict forgetting since they identify sets of student beliefs which must change (i.e. be "forgotten") after new conflicting information arises. Unfortunately, not all forgetting results from conflict resolution, so other principles must apply, perhaps based on theories of how learners learn (e.g. after general rules have been induced from specific instances of the rules, then the specific instances are often discarded). Sometimes it is possible to directly discern certain kinds of forgetting as reflected in learner behaviour, for example when a learner who used to be able to write recursive functions is no longer able to Most forgetting, though, won't be so happily observable or deducible. The truth maintenance system that must track forgetting won't even have the advantages of being able to use the current instructional goals of the learner and tutor that can help it when the student is learning new things. When a learner forgets something, it happens as a background activity, unconnected to the current goals of the tutor or learner. Even if such forgetting can be recognized, how does forgetting impact conflict detection and resolution? After a learner has forgotten some key concept, certain things may now be in conflict which weren't before, and vice versa. Current truth maintenance systems only deal in accretion, not deletion, so this promises to be a tricky problem as well, but an interesting one. Forgetting is one of the most neglected problems in artificial intelligence, but one which must ultimately be dealt with in student modelling.

Any progress in tackling these difficult problems would certainly be useful to AI more generally, thus proving that student modelling is, indeed, a crucible for AI research. But, by any lights, student model maintenance is clearly a very hard thing to do, when the attempt is to track the internal knowledge shifts of the learner. The difficulty of clearly understanding a learner's behaviour, combined with the extreme complexity of the shifting knowledge structures in a learner's mind, would seem to challenge the current capabilities of truth maintenance techniques rather too severely. If, on the other hand, we move outside the brackets to extend truth maintenance techniques to work at the level of the tutor's beliefs about the learner, progress may be possible. Hopefully, techniques that are useful outside the brackets can then be transferred to some extent inside the brackets.

This is precisely what Huang (1991) has done. Huang has developed a three part student model maintenance system, largely focussed on maintaining the tutor's own beliefs about the learner rather than the learner's internal beliefs. At its core is a working memory that contains a subset of beliefs the tutor has about the learner, those beliefs that are currently in focus (Huang et al., 1991a). As Huang and McCalla (1992) have shown it is reasonable to assume that such focus sets occur naturally in an instructional environment where the learner and tutoring system agree to concentrate on some subpart of the domain. The consistency of this working memory is maintained by a truth maintenance system in the style of deKleer's (1986) ATMS, although Huang's system only produces minimal changes to the working memory, rather than identifying all possible conflict sets (see Huang et 1991b). The working memory, however, is not the entire set of beliefs the tutor has about the learner. There are many other beliefs stored in the second component of Huang's architecture: the long term memory. Huang et al. (1991a) provide efficient algorithms for moving sets of beliefs (so-called "frames of mind") from long term memory into working memory and for moving them back. When beliefs move into short term memory from long term memory, they must be analyzed by Huang's truth maintenance system, both for internal consistency and for consistency with other beliefs already in working memory. When they move from working memory into long term memory, a quick check is made for obvious inconsistencies with other beliefs in long term memory, but no complete analysis is made. The final component of Huang's architecture is called the DPN: the data package network (see Huang et al., 1991b). The DPN stores stereotypical beliefs that different kinds of learners typically have at various stages of their learning.

The DPN can feed sets of default beliefs into the working memory, according to the stage the learner seems to be at, and can withdraw and replace these beliefs with new defaults as the learner is perceived to be evolving. Using the DPN, Huang's student model maintenance system can still reason about learners even in the absence of direct information by drawing on typical behaviour patterns.

Huang's approach illustrates a number of important ways in which investigations into student modelling, even working outside the brackets, have stretched traditional notions of knowledge representation and reasoning. First, there is no pretense that a knowledge base is consistent or complete. In fact, Huang's system has several levels of consistency. The long term memory need not be particularly consistent at all, with just occasional partial checks on consistency being carried out. The working memory is consistent to the level of an ATMS, i.e. inconsistencies are removed only if there are explicit justifications showing how one disbelieved proposition is connected to another which should now also be disbelieved. Finally, the beliefs in a stereotyped package in the DPN are consistent in the sense that learners can actually manifest this package of beliefs, but the beliefs are not necessarily logically consistent in any way. A second way in which student modelling has encouraged Huang to new insights is in his extensions of ATMS to provide a notion of minimality, based on the idea that learners exhibit cognitive economy, i.e. they change as little as possible when they see a conflict. Third, Huang has been both compelled and stimulated to deal with the idea of focus. He has been compelled by the fact that in a system as big as an ITS, an ATMS would not efficiently be able to maintain all beliefs. He has been stimulated to deal with focus by the fact that focussing makes sense in ITS, where often an instructional plan is driving the interaction and content items relevant to the current plan step can form the basis for the focussed beliefs in Huang's system (see Huang and McCalla, 1992). Finally, Huang has tried to represent different styles of knowledge, from observed beliefs to stereotypical beliefs, in one package because student modelling requires a combination of reasoning styles.

Even working outside the brackets, Huang has made progress on student modelling. In fact, it seems reasonable that some of his approaches (e.g. focussing) may prove to be useful inside the brackets too. Despite Huang's progress, however, a sophisticated student model maintenance system with most of the capabilities discussed in this section is a long way off, if it is even possible to achieve at all. In the short term, practical tutoring systems will have to content themselves with fairly straightforward belief revision of the kind Huang suggests, overlaid on knowledge structures akin to Elsom-Cook's bounded student models or our granularity hierarchies.

5   Other Issues in Student Modelling

There are many other issues inherent in student modelling for intelligent tutoring systems, some of them touched upon in the discussion so far, and some brand new. How can the knowledge in a student model be generated as needed rather than prestored and just overlayed? Can a principled theory of misconceptions be devised? Are such theories inherently domain specific (like BUGGY (Brown and Burton, 1978)) or are there domain independent theories of misconceptions which take into account possible general principles of reasoning and mal-reasoning (e.g. Repair Theory (Brown and VanLehn, 1980))? What is the relationship between domain knowledge and more general problem solving skills? Are there theories of cognition that can be used to explain the interconnections between a learner's mental models, strategies, and surface behaviour? How can an episodic record of student history best be acquired and organized? How can the fact that a learner's knowledge is necessarily incomplete and inconsistent, not to mention that a tutor's knowledge of the learner is normally also incomplete and inconsistent, be dealt with if the "neat AI" goal of having a principled (i.e. formal) approach to student modelling is to be retained?

Many of these issues are general to AI, although they are particularly poignant for ITS. Research into any of them may well suggest whole new approaches to knowledge representation and innovative solutions to old problems, and conversely may necessitate the marshalling of many AI techniques to resolve them. This illustrates once again the synergy between ITS and AI.

6   Conclusions

In this paper I have argued that student modelling stimulates innovative research. To illustrate this point, I have discussed how the demands of student modelling have led to new ideas in granularity-based knowledge representation and to interesting extensions to standard truth maintenance paradigms. When trying to model a learner, issues such as consistency, completeness, and certainty become irrelevant; the learner is a moving target, constantly changing, very hard to understand. Any knowledge representation scheme that is meant to model a learner's knowledge, therefore, has to be flexible, adaptable, and robust in the face of these uncertainties. It is my contention that these uncertainties are also realistic for all human behaviour, not just the behaviour of learners in an intelligent learning environment. Thus, consistency, completeness and certainty are not particularly prevelant features of human knowledge. The new ideas developed in student modelling, therefore, should readily apply to artificial intelligence and cognitive science more generally.

These new ideas also help to make student modelling practical. Granularity, for example, allows a tutoring system to diagnose learner behaviour effectively, even when learners stray far from predicted paths. Focussing allows a truth maintenance system to efficiently maintain a subset of beliefs relevant to current learning goals, rather than having to monitor an intractably large knowledge base. Stereotypes provide a student modelling system with default knowledge of typical learners even in the absence of specific knowledge of a particular learner. Each of these techniques achieves success by removing the requirement for perfect knowledge of the learner. As long as the tutoring system is able to keep the learner engaged and thinking about his or her problems, then it is succeeding. As with human tutors, a tutoring system will enhance the chances of keeping the learner engaged by individualizing the feedback, by trying to adapt to the learner, even if it is sometimes less than perfect. The measure of the success of the student modelling system, therefore, is not total accuracy, but usefulness.

There are many exciting developments that lie ahead for student modelling research. Most importantly, I think, is the critical need for the student modelling enterprise to move into the arena of investigating social context. Research into situated learning (Clancey 1992) provides convincing evidence of the importance of social context to learning. Learners are more likely to understand new ideas if the ideas are grounded in things they know about from their everyday life. They are more likely to be motivated to learn if the learning environment allows them to achieve their own goals.

The importance of social context has led many researchers investigating situated cognition to reject the need for, even the possibility of, student modelling. These researchers, however, take a very narrow view of student modelling, tending to think of it as an enterprise that tries to categorize learners in terms of how their knowledge does or does not instantiate expert knowledge. In contrast, I see student modelling as trying to categorize learners in terms of the factors that influence their learning. These factors certainly include their level of domain knowledge and how this level compares to expert knowledge, but not just this. There is nothing that precludes student modelling from factoring in the motivations of learners, their social situation, their goals, their emotional states, etc.

It is, of course, a difficult problem to be able to infer these things, but I see no reason in principle that it couldn't be done. Granularity, focus, stereotype reasoning, and no doubt other techniques might all be brought to bear on the difficult problem of understanding these factors, if not completely for a given learner, at least to some degree. By trying to extend student modelling to deal with these new factors, whole new areas of research may open up for artificial intelligence, which may ultimately lead to representation-oriented investigations of social context. If such investigations succeed, or even partially so, this might remove what I see as an artificial polarization between situated cognition research and artificial intelligence research. This would allow the strengths of each perspective to come together to create a powerful new theory of human cognition, founded on a representation-based approach to modelling social context. Such a grandiose outcome would indeed prove that student modelling is a crucible for research.

Acknowledgements

I would especially like to thank Jim Greer who for the past six years has been a valuable colleague and a significant contributor to the ARIES Laboratory as well as to all of the ideas that have arisen in our research. Particular thanks also go to all of our graduate students and research assistants, most especially to Xueming Huang, Bryce Barrie, Paul Pospisil, Mary Mark, Dinesh Gadwal, Jason Neudorf, and Randy Coulman who have worked on research directly pertaining to the issues discussed in this paper. Of course, any bizarre intepretations put on these ideas by me in this paper are my own responsibility, not theirs! Finally, I would like to acknowledge the financial support of Canada's Natural Sciences and Engineering Research Council and the University of Saskatchewan.

References

(NATO ASI Series F volumes are indicated by F and volume number)

Anderson, J. and Reiser, B. (1985) The LISP tutor. Byte, 10(4), 159-175

Bhuiyan, S.H., Greer, J.E., and McCalla, G.I. (1989) Mental models of recursion and their use in the SCENT programming advisor. Proceedings of the 1989 Conference on Knowledge-Based Computer Systems, pp. 135-144. Bombay: Narosa

Bhuiyan, S.H., Greer, J.E., and McCalla, G.I. (1991) Characterizing, rationalizing, and reifying mental models of recursion. Proceedings of the 13th Annual Meeting of the Cognitive Science Society, Chicago, IL, pp. 120-125

Brown, J.S. (1990) Toward a new epistemology for learning. In: C. Frasson and G. Gauthier (eds.), pp. 266-282. New Jersey: Ablex

Brown, J.S. and Burton, R. (1978) Diagnostic models for procedural bugs in basic mathematical skills. Cognitive Science Journal 2, 155-191

Brown, J.S. and VanLehn (1980) Repair theory: a generative theory of bugs in procedural skills. Cognitive Science Journal 4, 379-426

Clancey, W.J. (1992) New perspectives on cognition and instructional technology. In F91

de Kleer, J. (1986) An assumption-based truth maintenance system. Artificial Intelligence Journal 28, 127-162

Doyle, J. (1979): A truth maintenance system. Artificial Intelligence Journal 12, 231-272

Elsom-Cook, M.T. (1988) Guided discovery tutoring and bounded user modelling. In: J. Self (ed.) Artificial intelligence and human learning: intelligent computer-aided instruction. London: Chapman and Hall, 165-178

Goldstein, I. (1982) The genetic graph: a representation for the evolution of procedural knowledge. In: D. Sleeman and J. Brown (eds.) Intelligent tutoring systems. London: Academic Press, pp. 51-78

Greer, J.E. and McCalla, G.I. (eds.) (1994) Student modelling: the key to individualized knowledge-based instruction. F125

164   G. I. McCalla

Gutwin, C. and McCalla, G.I. (1992) Would I lie to you? Modelling misrepresentation and context in dialogue. Proceedings of the 30th Conference of the Association for Computational Linguistics, U. of Delaware, Newark, Delaware, pp. 152-158

Hobbs, J. (1985) Granularity. Proceedings of the 9th International Conference on Artificial Intelligence, Los Angeles, CA, pp.432-435

Huang, X. (1991) Updating belief systems: minimal change, focus of attention, and stereotypes. Ph.D. Thesis, U. of Saskatchewan: Department of Computational Science

Huang, X., McCalla, G.I. and Neufeld, E. (1991a) Using attention in belief revision. Proceedings of the American Association for Artificial Intelligence Conference, Anaheim, CA, pp. 275-280

Huang, X., McCalla, G.I., Greer, J.E. and Neufeld, E. (1991b) Revising deductive knowledge and stereotypical knowledge in a student model. User Modelling and User Adapted Interaction Journal 1(1), 87-115

Huang, X. and McCalla, G.I. (1992) Instructional planning using focus of attention. Proceedings of the 2nd International Conference on Intelligent Tutoring Systems. Montreal, Canada. Lecture Notes in Computer Science, Vol. 608, pp. 443-450. Berlin: Springer-Verlag

Mackworth, A. and Havens, W.S. (1981) Structuring domain knowledge for visual perception. Proceedings of the 7th International Joint Conference on Artificial Intelligence. Vancouver, Canada, pp. 625-627

McCalla, G.I. (1992) The centrality of student modelling to intelligent tutoring. In F91, pp. 107-134

McCalla, G.I., Greer, J.E., Barrie, J.B., and Pospisil, P. (1992): Granularity hierarchies. International Journal of Computers and Mathematics, Special Issue on Semantic Networks, 23, pp. 2-5, pp. 363-376

Self, J. (1992) Are theories of diagnosis applicable to cognitive diagnosis? Third international workshop on user modelling, Schloss Dagstuhl, Germany

10

Diagnosis and Cognitive Support in Microworlds and Intelligent Tutoring Systems: Conceptualization, Realization, and Application to Real-World Learning Environments

David L. Ferguson

Departments of Technology and Society, and Applied Mathematics and Statistics
State University of New York, Stony Brook, New York 11794-2250

Abstract: This paper explores methods for both diagnosing students' under-standing and cognitively supporting students' problem solving efforts in micro-world and intelligent tutoring learning environments. Special attention is given to how such aids for diagnosis and cognitive support are implemented within the microworld and intelligent system tools and the learning contexts in which such tools are applied. Student models are viewed as a subset of the larger set of meth-ods used for diagnosis and cognitive support. Several examples, drawn from micro-worlds, intelligent tutoring systems and hybrid systems, illustrate these ideas in environments aimed at helping students develop scientific explanations (theories) or solve novel problems. The paper considers how developers of microworlds and intelligent tutoring environments might mutually benefit from each other's work, and how all such development must co-evolve with the total learning environ-ments in which the tools will be deployed.

Keywords: Microworlds, intelligent tutoring systems, diagnosis, cognitive support, student models, problem solving, learning environments

1 Introduction

Understanding students' problem solving behavior and offering support for students as they struggle with concepts and problems are critical elements of computer-based problem solving environments. Developers of microworlds, advocating great freedom for the learner, have often seen themselves at opposite ends of the learner-control spectrum from developers of intelligent tutoring systems. However, devel-opers of each type of environment (microworld or intelligent tutor) acknowledge that the environment must offer diagnostic support and make pedagogical decisions.

This paper has three major aims: 1) to clarify the characteristics of microworlds and intelligent tutoring systems, 2) to show how diagnosis and cognitive support are conceptualized and implemented in microworlds and intelligent tutoring systems, and 3) to suggest how developers of microworlds and intelligent tutoring systems might benefit from each other's work. Specific examples of current environments are used to illustrate the concepts.

2. Features of Microworlds

The term "microworld" has been used by a number of researchers to describe computer-based learning environments that are highly visual and geared to open-ended problem solving. Microworlds have several characteristics (Ferguson, 1992):

a) Microworlds are self-contained environments. (The environment is in some sense closed. In Logo, the tools necessary to design are available in the environment. Similarly, the microworld Cabre-Geometre provides a wide range of geometric tools for students to study problems in elementary Euclidean geometry.)

b) Microworlds are constrained environments. (For example, the universe of discourse in Logo is constrained by the language itself.)

c) Microworlds provide opportunities for multiple views and representations of real phenomena. (In Logo, the computer program and graphical representations give different views.)

d) Microworlds make it easy to learn by constructing objects. (Students learn by "designing".)

e) Microworlds are rich in the variety of questions that can be posed and potential solutions that can be explored.

f) Microworlds make it easy to learn via debugging activities.

Many developers of microworlds were influenced by Seymour Papert's *Mindstorms* (Papert, 1980). We can capture some of the meaning of "microworld" from Papert's own writing.

"We must ask why some learning takes place so early and spontaneously while some is delayed many years or does not happen at all without deliberately imposed formal instruction." "If we really look at the "child as builder,' we are on our way to an answer. All builders need materials to build with. Where I am at variance with Piaget is in the role I attribute to the surrounding cultures as a source of these materials. In some cases the culture supplies them in abundance, thus

facilitating constructive Piagetian learning. For example, the fact that so many things (knives and forks, mothers and fathers, shoes and socks) come in pairs is a "material" for the construction of an intuitive sense of number. But in many cases where Piaget would explain the slower development of a particular concept by its greater complexity or formality, I see the initial factor as the relative poverty of the culture in those materials that would make the concept simple and concrete. In yet other cases the culture may provide materials but block their use. In the case of formal mathematics, there is both a shortage of formal materials and a cultural block as well. The mathophobia endemic in contemporary culture blocks many people from learning anything they recognize as 'math,' although they may have no trouble with mathematical knowledge they do not perceive as such."
(from *Mindstorms* by Seymour Papert, Basic Books, New York, 1980, page 7)

3   Features of Intelligent Tutoring Systems

A good human tutor has many capabilities:

- Subject-domain expertise,
- Ability to diagnose student's errors (and formulate appropriate models of the students' state of knowledge),
- Ability to use a variety of instructional approaches and adapt the approaches to the needs of the learner,
- Ability to present captivating tasks and sustain the interest of the learner over extended periods.

Over the last fifteen or twenty years there has been considerable effort to develop computer-based tutoring systems that exhibit "human-like" tutoring expertise. Efforts in this direction, regardless of the extent to which they actually meet their goals, have generally been described as intelligent tutoring systems.

Intelligent tutoring systems are usually characterized by a subset of the following features:

a) They attempt to teach the solving of problems that do not readily lend themselves to simple algorithmic solutions. (The task might be electronic trouble shooting, solving symbolic integration problems, solving certain classes of physics problems, solving algebra word problems, doing proofs in elementary Euclidean geometry, or tackling other problems in mathematics, science, engineering or a variety of other fields.)

b) The system accepts and processes a wider range of "natural language" input than is normally expected of computing systems.

c) The system is able to use certain rules to make inferences or deductions, and hence arrives at new knowledge that may influence its subsequent behavior.

d) The system exploits instructional strategies that partly mimic those of good human tutors. (Such abilities might include the capability to diagnose the students' conceptions about a particular content area, and offer certain remedies.)

e) The system possesses some learning capabilities. (Notice that this is implied, in part, by item c above.)

f) The system possesses a mechanism for accepting a wide range of symbolic representations (graphs and diagrams, informal language, formal symbolic language, etc.) and "reasoning" about such information.

g) The system possesses a feature of transparency that allows the user to "ask" the system for the "line of reasoning" that it used to arrive at a particular conclusion, permitting the system to give a direct and detailed response to the query.

h) The system possesses a feature of extensibility that permits major changes to be made in the models that the system uses to solve problems, and in the instructional strategies that the system employs to aid students.

Much of the research in the development of intelligent tutoring systems has concentrated on four major areas of these systems: domain expertise, student model, pedagogical model and user interface. The "student model" issue has presented a tremendous challenge to researchers (e.g. see McCalla, this volume).

4   Diagnosis and Cognitive Support in Microworlds: The Externalization of Knowledge

In this section, we discuss diagnosis of students' behavior and cognitive support for students within the context of microworld-supported learning environments Diagnosis and cognitive support in microworld-supported learning environments are greatly dependent on indirect feedback (e.g. graphics outcomes of students' action in Logo) and peer or teacher support.

Microworlds utilize the power of the "externalization of knowledge". Through such externalization knowledge becomes inspectable by the learner and by others who seek to aid the learner. Microworld-supported learning environments use a wide range of methods to aid in diagnosis and cognitive support. We now describe many of these strategies:

a) appropriate constructs in the environments (i.e. appropriates tools such as specialized languages or manipulatable diagrams),

b) tasks appropriate for specific concept development,

c) direct manipulation (special objects may be transformed by the learner via a set of permissible transformation "rules"),

d) effective feedback (feedback is indirect: the learner and those who support the learner are expected to infer something about the learner's conceptions by inspecting the feeback),

e) visualization,

f) multiple representations,

g) constructivist learning ("constructivism" takes the view that meaning is shaped by the learner's attempt to relate new experiences to her/his existing conceptualization of the world; learning is viewed as a dynamic process that transforms an old state of knowledge into a new state of knowledge which incorporates or challenges a new set of experiences),

h) constructionist learning (Seymour Papert has used the term "constructionism" to refer to the explicit use of tools that are external to the learner in the fostering of constructivist learning),

i) constrained environments,

j) training wheels environments (a sequencing of open-ended learning environments where the complexity of environments is increased by allowing the user to manipulate a greater number of parameters or to create more complex entities with the same set of parameters),

k) discourse-sensitive manipulations (the manipulation of objects may have a greater impact on learning when such manipulation draws on metaphors from objects in the real world and when the manipulation has a natural correspondence to ordinary spoken or written language),

l) external (inspectable) knowledge,

m) learner, peer, and teacher support.

5   Diagnosis and Cognitive Support in Intelligent Tutoring Systems: In Search of Student Models

One of the more complex problems in the design of intelligent tutoring systems is the development of models of the students' conceptualizations. Ideally, such models must be dynamic and capture the temporal status of the student's understanding as various understandings evolve. In practice, it is extremely difficult to develop models that reflect students' understanding. Hence, the designer is often in a weak position in the effort to offer cognitive support.

Researcher's conceptualizations of the student model have evolved over the relatively short research history of intelligent tutoring systems. VanLehn described the three characterizing dimensions of student models: bandwidth, knowledge type,

and student-expert difference (VanLehn, 1988). He clarifies the values along each
of these dimensions as follows:

1. Bandwidth – How much of the student's activity is available to the diagnostic
   program?
   a) Mental states – All the activity, both physical and mental, is available.
   b) Intermediate states – All the observable, physical activity is available.
   c) Final states – Only the final state – the answer – is available.
2. Knowledge Type – What is the type of subject matter knowledge?
   a) Flat procedural – Procedural knowledge without subgoaling.
   b) Hierarchical procedural – Procedural knowledge with subgoals.
   c) Declarative.
3. Student-Expert Difference – How does the student model differ from the expert
   model?
   a) Overlay – Some items in the expert model are missing.
   b) Bug library – In addition to missing knowledge, the student model may
   have incorrect "buggy" knowledge. The bugs come from a predefined library.
   c) Bug part library – Bugs are assembled dynamically to fit the student's
   behavior.

General theories of cognition may offer little insight into the learning of
specific subject matter and may be inadequate to support the building of detailed
student models. In recent work, VanLehn argues that "task specific theories of
learning" are needed (VanLehn, 1992). He describes a set of integrated computer-
based tools for analyzing student data and building models that account for the
data.

McCalla presents three major ideas to make student modeling practical
(McCalla, this volume): "*Granularity*, for example, allows a tutoring system to
diagnose learner behavior effectively, even when learners stray far from predicted
paths. *Focusing* allows a truth maintenance system to efficiently maintain a
subset of beliefs relevant to current learning goals, rather than having to monitor
an intractably large knowledge base. *Stereotypes* provide a student modeling
system with default knowledge of typical learners even in the absence of specific
knowledge of a particular learner. Each of these techniques achieves success by
removing the requirement for perfect knowledge of the learner."

6 The Principled Use of Microworlds and the Flexible Use of Intelligent Tutors

Developers (and users) of microworlds and intelligent tutoring systems have much
to learn from each other. The principled design of a microworld for a particular
subject domain (geometry, thermodynamics, etc.) demands that the developer
construct "theories" to explain students' conceptions and formulate "theories" of

instructional processes that might foster changes in students' understanding. Such theories become the cornerstones for curriculum development and general pedagogical practice. In order to design flexible intelligent tutoring systems, developers must recognize that intelligent tutoring systems must leverage the human diagnostic support (learner, peers and teachers) and other instructional support that may be available in the learning environment. Intelligent tutoring systems must capitalize on the intelligence of tutees, especially the tutees' ability to do a great deal of self-diagnosis when given powerful indirect feedback systems. This view of developers of microworlds and developers of intelligent tutoring systems suggests that the "two camps" might greatly from each other's work.

Recent work in the development/use of microworlds and intelligent tutoring systems suggests a more synergistic relationship between the two camps. Linn's work, incorporating a microworld--type environment and microcomputer-based laboratory tools, demonstrates how an understanding of "students' action knowledge" (actions generated by students in the real world), "students' intuitive conceptions", and "scientific ideas" yields a perspective on integrated scientific knowledge that helps middle school students to understand thermodynamics (Linn, 1993). Similarly, Thornton's pioneering studies of the use of microcomputer-based laboratories in the teaching of physics, has shown how a principled approach to the teaching of motion deepens students' understanding and problem solving ability (Thornton, 1992). Henderson and Ferguson use a hybrid system (mix of microworld and intelligent tutoring approach) to enhance college freshmen's abilities to discover and express algorithms in a Foundations of Computer Science course (Henderson and Ferguson, 1992). Recent work of C. Laborde, J. Laborde, and others demonstrates how fundamental ideas about how students learn geometry are producing a hybrid (integrated microworld and intelligent tutoring) environment that builds on Cabri-Géomètre (C. Laborde and J. Laborde, 1992). All of these efforts point to potential mutual benefits for workers in microworld and intelligent tutoring systems.

7   Conclusion

The greatest power of computer-based learning activities is derived when such activities are creatively and effectively situated (cognitively and affectively) in purpose-driven, rich and flexible learning environment. Developers of microworlds and intelligent tutoring systems might very well benefit from each other's work. The student modeling issue that is so critical to work in intelligent tutoring systems might better be viewed as a subset of the more general set of methods for doing diagnosis and supporting students as they develop understanding and solve problems. The principled development of microworlds demands that the developer construct "theories" about the learner and the process of "instruction". As more hybrid systems emerge, developers will be forced to consider the appropriate utilization of each approach (microworld or intelligent tutoring) for specific purposes. All such development must co-evolve with the total learning environments in which the tools will be deployed.

References

(NATO ASI Series F volumes are indicated by F and volume number)

Ferguson, D.L. (1992) Computers in teaching and learning: An assessment of current practices and suggestions for future directions. In F96

Henderson, P.B. and Ferguson, D.L. (1992) Using structure as a design tool in algorithmic problem solving. In: D.P. Balestri, S.C. Ehrmann and D.L. Ferguson (eds.) Learning to design, designing to learn: Using technology to transform the curriculum. Washington, DC: Taylor & Francis

Laborde, C. and Laborde, J. (1992) Problem solving in geometry: From microworlds to intelligent tutoring systems. In F89

McCalla, G. (1996) Student modeling: A crucible for research. In this volume

Papert, S. (1980) Mindstorms. New York: Basic Books

Thornton, R.K. (1992) Enhancing and evaluating students' learning of motion concepts. In F86

VanLehn, K. (1988) Student modeling. In: M.C. Polson and J.J. Richardson (eds.) Foundations of intelligent tutoring systems. Hillsdale, NJ: Lawrence Erlbaum

VanLehn, K. (1992) A workbench for discovering task specific theories of learning. In F96

11

Instructional Design for Computer Based Learning Environments

Martial Vivet

LIUM, BP 535, Université du Maine, F-72017 Le Mans cedex, France

Abstract: The paper considers methodological issues in the design of Computer Based Learning Environments. The methodological ideas we have developed suggest that one should start the design with a precise specification of a scenario of use including precise roles for the learner(s) and teacher(s). We show the limitations of ITS prototypes as they have so far been built and explain why such systems are not really used. We also develop the idea that we need knowledge dealing with the kind of interventions the teacher can make adressing the balance between tutored phases and free discovery time for the learner. Finally, we claim that not enough has yet been done to take into account results from the educational sciences and that we should be more precise in our specification of the connection between learning theory and design for a given system. A few possible paths for the future are suggested: the need for more prototypes to help researchers learn from effective large scale use, and also the need for more formalisation with precise linkage to learning theories.

Keywords: Design, methodology, computer based learning environments, interactive learning environments, intelligent tutoring systems

1 Introduction

We shall consider both fundamental and methodological issues in the design of Computer Based Learning Environments using knowledge based techniques.

A basic idea that we are currently developing about methodology is to start the design with a precise specification of a scenario of use with precise tasks and roles for the learner(s) and teacher(s). Only from there can a relevant knowledge analysis, in terms of knowledge in the domain to be taught, student modelling, interaction and dialogue management, be given. We claim that this knowledge analysis depends on the initial step. As an example, the production of explanations is only useful if the explanations are relevant (at the right time, with the right content in the right form!), which depends on their use, and it is from this use that the explanation process gains some feedback about the knowledge representation. We then show some limits of the prototypes developed in the Intelligent Tutoring Systems (ITS) community as they have been built until now and how this can explain why such systems are not really used.

Merging ideas from constructivism and micro-worlds, we also develop the idea that we need knowledge dealing with the role of the teacher [38], and the kind of interventions he/she can make. These ideas lead us to discuss the balance between, on the one hand tutored phases and on the other, time for the learner allowing periods of free discovery.

As a conclusion, we claim that not enough has yet been done to take into account results from educational sciences. For example, we underline the need for a more precise specification of the connection between learning theory design for a given system. This appears to be the right way to validate the foundations of a system at the theoretical level. Finally we consider now this work might progress: the need for more prototypes, which are used in real-world learning contexts, to help researchers learn from effective large scale use of such training systems, and also the need for more formalisation with precise and complete linkage to learning theories and educational sciences.

2   ITSs, ILEs, SLEs, ALEs

The Intelligent Tutoring Systems (ITS) concept came from merging Computer Based Teaching (CBT) systems ideas with knowledge based systems (KBS) design in AI. The main goal was to produce adaptability, flexibility, and individualisation in the communication process with the student. This approach to designing systems can be characterised as starting from the knowledge representation in the domain to be taught. A major development in representation techniques has been imposed by the need for explanation.

The Interactive Learning Environments (ILE) concept emerged from the use of hypertext and hypermedia [7] techniques to design computer based learning material. The main goal was to give the learner the possibility for free browsing among knowledge items. The approach to designing such systems can be characterised by a strong structuring of the knowledge in the domain to be learnt (finding basic knowledge items and eliciting the relations between them).

Situated Learning Environments (SLE) introduced ideas around the context of use of knowledge [16, 43–45]. SLEs are more concerned with learning theories based on constructivism. The approach to designing such systems is characterised by the design of micro-worlds [11] where learning takes place essentially from problem solving activities.

Adaptive Learning Environments (ALEs) [15] are a step forward and encompass ITSs with AI techniques, ILEs and SLEs.

This development can be characterised by an increased consideration of learning [9, 47] (rather than teaching), with admission of the need for a balance between tutoring and free access to knowledge items and the need to consider the role of the learner within the environment.

ALE design pays more attention to learning theories and is an experiment in merging several techniques for implementation from the computer science point of view.

Remark: CBT systems are very often blamed for allowing only one pedagogical strategy, but it is also clear that generally knowledge based learning systems don't show enough variety and richness in pedagogical strategies either. Promises made are still only hoped for even if approaches to make available explicit strategies given in a declarative way are under development [20-22, 34].

3   Didacticians, Didactics and Learning Situations

Didacticians (following the French meaning) are concerned with the process of acquisition of the contents of the sciences; they examine relations between teaching content and learning it and try to produce recommendations at a detailed enough level to allow the design of tools for the teacher. Their work has a strong connection with epistemology.

Didactics can be characterised by the fundamental concepts it manipulates: didactical situations (theory of situations by Brousseau [4, 5]), didactical contract, didactical transposition, learner's beliefs, the problem of transfer and evaluation.

The reconstruction of knowledge in the learner's mind from a chain of cognitive activities like action – formulation – validation – institutionalisation is a major topic of interest.

It is clear that we need to take more account of learning theories [9, 47] to lay stronger foundations for the systems which are designing. A good result would be achieved if we can arrive at a practical validation of these theories.

4.   Design for Computer Based Learning Environments

4.1 Preliminary Remarks

Most ITSs (even existing ones! [10, 23, 38]) are not effectively used. ILEs are used but we know very little about what is effectively learnt with them. SLEs can be more effective from the learning process point of view but need very efficient teachers who are in short supply.

It appears then that the early design of these systems is generally not done in the right way. For example most ITSs have been designed from an analysis of the knowledge and reasoning process in the domain to be taught. It is only once the system is built that its use is explored. The teacher has no pre-defined role with regard to the use of the system. This probably comes from aspects of the old dream of automatic teaching machines which are still surfacing.

4.2 Methodological Specifications for Design

4.2.1 Design Principles

Our proposal is to *start from the "learning situation"* with a systematic approach. This first gives a specification of use; a clear description of the context and activity of the learner and examples of interaction scenarios with the learner in the "milieu" [4]. The design must be centred on learner's tasks. The computer software to be written is only a component in this milieu, with relations to be carefully studied and specified with other components. This imposes precision in the description of the role of the teacher, of peers in the milieu, and different pedagogical functions of the software (availability of ITS, ILE, SLE aspects, cooperation and communication between these aspects). The most important thing, right at the start, is the specification of the cooperation process between participants (including the software and the learner) in this active "milieu". This is the only way to build systems which arrive in the right place at the right time.

Next, a description of the dialogue managment process (the kind of dialogue model to be used: negotiation for example [1]), of the interaction and of the communication interface needs to be given. The ergonomics of the interface should be designed at this stage.

The design of the evaluation process which will be used ought to be done early too. The technical evaluation (evaluation of tools, ergonomics of screens, quality of interaction) must not hide the pedagogical evaluation requirements (evaluation of the learning process available, learning process results).

The definition of a developers team with people having different backgrounds is very important: AI, computer science, didacticians and teachers are all required.

4.2.2 Architect Metaphor

The designer has a very important role which goes further than usual project management tasks. He is acting as an architect, designing an environment for learning, somewhere for the intellectual life of the learner. He has to design blocks of technical resources with a clear specification of the modes of access to each, to make paths available (main roads where everybody should go at least once and small streets to allow personal free discovery), meeting points to meet others (teacher, peers), public places where the learner's activities can be seen or a private place where the personal activity of the learner is hidden from others. This architect doesn't need to have specific knowledge of the domain to be taught. His role is to feel and conduct like an orchestral conductor.

4.3 The Spiral Process

This model [3] for design and development is based on progressive refinements, evaluation and re-evaluation phases and enlargement of user classes. First design is done from the learning situation class specifications and is improved taking into account remarks from users. Users are expert teachers in the domain and very naïve learners, evaluation being done by didacticians. Later on, users who are normal teachers and students are questioned. Main tasks include prototyping (translation of specification in terms of constraints on the design and the technical realisation), technical validation of software, ergonomic validation (This is done in two stages: in the design phase with developers and in a learning situation with students), pedagogical validation of the reference learning situation and validation of the feasability of the roles of teachers to allow generalisation of use.

4.4 Design of Shells for the Development of Applications

4.4.1 Knowledge Representation and Explanation

A number of tools are under development [12, 13, 22, 34] to help the production of computer based learning material. Basic ideas now consist in merging know-how from Computer Assisted Learning (CAL) authoring systems and knowledge based systems shell design.

With the KBS approach, a lot of work has been done to construct knowledge representations allowing explanations [35]. The problem of the production of explanation has required a large amount of work at the technical level. But fundamental questions have not yet been answered: how to produce timely explanations? What is a relevant explanation in a given situation? In a didactical situation we are deeply concerned by the "right explanation at the right moment". These aspects should be combined with the fact that often the best way to help the learner learn is to remove the possibility of any explanation at all.

4.4.2 Pedagogical Expertise

There is a specific focus these days to produce flexible tutorial strategies (the best so far done is to have an explicit declarative representation of pedagogical expertise in pedagogical rules). Some attempts are promising even if the challenge is still hard to surmount. Models are available [12, 20–22, 34] which have proved that the path is possible. This process is very difficult and a major problem is that a large part of pedagogical expertise is embedded in the interaction process [32]. For example, it is clear that an important aspect of the pedagogical expertise of a teacher of foreign languages is his/her capacity to interact directly, in real time, with the learner using a language level (a sort of intermediate language step as in MARPLE [32]) compatible with the learner's.

4.5 Directions for Future Work

It seems clear now that we need more work and better results on a number of different aspects:

i) design of a language to specify situations, to describe learning situations and elicit constraints on tools and roles of participants. This formalisation step can produce tools to help the early stages in the design process but also during the evaluation phases. Such an intermediate language, expressing pedagogical problems, could help didacticians to express their knowledge.

ii) From interaction to dialogue management: dialogue management must be improved. It would be interesting to build systems conscious of the kind of dialogue engaged in with the learner at a certain pedagogical phase. Most of the work done so far concerns negotiation [1]. It would be convenient to articulate negotiation phases with phases where the learner works freely (ILE approaches) with only sporadic advice and with phases allowing more directive/imperative aspects. Difficulties arise from the need for specific knowledge to control the change between phases. The most promising approaches seem to be in the search for techniques allowing effective cooperation between the learner and his environment. The problem of student modelling is also still looking behind the global problem of communication and is itself only a part of the overall solution. We need modelling of the tasks engaged in by the learner, and modelling of intervention procedures by the system.

iii) Time management: the above remarks lead us to specify the need for effective research work based on time management in the learning situation. Very little research has yet been done, although work on pedagogical planning indirectly refers to it. If we consider teaching as a way to help a learner save time in his learning process we feel that time should be considered everywhere in this process. Timing is behind the planning of tasks and activities for/by the learner, and it is behind dialogue management (to allow coherence during the time between interactions with the learner). The clock of the computer which can refer to external time (inside and between sessions) does not necessarily fit with the internal clock of the learner or the internal clock of the teacher. A learning session always includes a lot of adjustments and compromises between these different clocks. Timing is an object to be studied in its own right in Computer Based Learning Environments design. Approaches for such research could start with the studies done in the educational sciences and take into account techniques developed in AI (generally around Computer Integrated Manufacturing applications development). Merging these ideas could lead to explicit management of time in a relevant way in CBLE.

iv) Production of systems effectively used: We need effective large scale use of systems under development to learn what happens to the learner and what happens to the teacher, the peers,... N. Major [21] tried to gather information with such an approach but we still need more. To do so, it is necessary to push the research work beyond the prototyping level to build well developed software (completeness, ergonomics of the interface, simplicity for use, largely distributed workstation criteria), and software robust enough to be used in a real context by a teacher needing no specific help from the developer! Methodologies to make correct observations of the learning process in such contexts need to be made more precise. The language for specifying the learning situation (cf. i) could help here a posteriori to allow correct descriptions of situations.

5 Example: ROBOTEACH

5.1 Goals

In our research group, P. Leroux is working in the area of control technology using micro-robotics environments. The approach is largely based on SLEs and has proved to be effective. However, these environments are limited by what we call "oversolicitation" of the teacher. The teacher is "oversolicited" because he/she plays different roles, particularly teaching programming concepts to program the movements of micro-robots, and helping the student to activate the micro-robots and to debug the programs.

ROBOTEACH is a system designed to allow the teaching of the basic ideas of driving robots. It is a flexible system which allows the creation of personalized learning sessions by the teacher. This means that the system manages information on the kind of interventions the teacher can make and gives the teacher the possibility of managing pedagogical parameters and adjusting them to needs detected for the student. For a student the pedagogical session corresponds to programmed access to different learning units. A particular unit is an automatic generator of driving programs from a micro-robot description made by the student. So, major functionalities of the system include:

– giving teachers the means to create pedagogical sessions in a micro-robotics environment.
– giving students help during a pedagogical session to understand basic concepts in learning how to drive robots.

5.2 System Description

The system consists of a generator with different screens to design the pedagogical session. The teacher specifies the chain of the learning units. The different units are: lesson, description, activating and debugging driving programs. For the lesson, the teacher specifies useful electronic cards and exercises, and for the description, activation and debugging units, he/she specifies the exercises (building or designing a micro-robot with one or several kinematics chains), the level of the student (beginner-1, beginner-2 or confirmed, which defines the kind of interactions between the system and the student and the means of description and activation), which electronic knowledge cards will be useful and the remediation tools available. Moreover the teacher plans his/her own intervention.

5.3 Implementation

ROBOTEACH is under development. We are using the PC-based hypertext system Toolbook to build the pedagogical session generator and to design the electronic cards.

6   Conclusion

The cooperation between AI researchers and didacticians can be very rich and is one way to gain the advantage of the latest information technology. This can lead to new possibilities, as long as we take care not to redo old approaches with new tools. The goal is not simulation of the blackboard!

More work is needed on these methodological aspects in the design of computer based learning environments. Firstly because this can help to build more efficient dialogue tools with researchers having different background, secondly to reach precise enough descriptions to allow effective pedagogical evaluation of the systems, thirdly to help in reducing costs, and to increase well qualified and available man power to develop such systems.

Acknowledgements

Most of the ideas in this paper emerged through discussions with colleagues at the Laboratoire d'Informatique de l'Université du Maine (LIUM) and within the national French group supported by the CNRS via PRC/IA - EIAO group and GR didactique. I would particularly like to thank Eric Bruillard who co-authored a paper in French from which parts of this paper are drawn. Discussions with N. Balacheff were also very influential. Nigel Major discussed the content and corrected the English language aspects; thank you!

References

(NATO ASI Series F volumes are indicated by F and volume number)

1. Baker, M. (1992) Negotiating goals in intelligent tutoring dialogues. In F91, pp. 229-255
2. Bessagnet, M.N., Nodenot, T., Gouarderes, G., Rigal, J.J. (1990) A new approach: courseware engineering. Proc. IFIP WCCE'90, North-Holland
3. Boehm, B.W. (1988) Understanding and controlling software costs. IEEE Transaction on Software Engineering 14(10)
4. Brousseau, G. (1988) Le contrat didactique: le milieu. RDM 9(3), 309-336, La pensée sauvage
5. Brousseau, G. (1986) Fondements et méthodes de la didactique des mathématique. RDM 7(2), 33-115, La pensée sauvage
6. Bruillard, E., Vivet, M.(1993) Concevoir des EIAO pour dessituations scolaires; approche méthodologique. RDM, N° spécial, to appear
7. Bruillard E., Weidenfeld G.(1990) Some examples of hypertext applications. In F67
8. Cauzinille E, Joab M., Mathieu J.(1989) NAIADE, a knowledge based system for explanation. Proc. 4th Int. Conf. AI&ED, Amsterdam, 24-26 May, pp. 42-46
9. Clancey W.J. (1992) New perspectives on cognition and instructional technology. In F91, pp. 3-14
10. Delozanne, E., Bruillard, E., Vivet, M. (1992) Mathematical learning environments: a French view point. In H. Nwana (ed.) Mathematical intelligent learning environments, pp. 14-32. Intellect books
11. Dickson, P., Heeter, C., Brade, K., Hohmann, L., Tabak, I., Weingrad, P. (1990) Exploratory multi-media environments. In F85, pp. 155-172
12. Futtersack, M., Labat, J.M. (1992) QUIZ: a distributed intelligent tutoring system. In: I. Tomek (ed.) Proc. ICCAL '92, Wolfville, 17-20 June, pp. 225-237. Springer-Verlag
13. Gouarderes, G. (1990) Le projet AGDI. TSI 9(5)
14. Jacoboni, P. (1992) Evaluation of pre-requisites. A milestone to evaluate technology in the educational process. Proc. ICTE, Paris, March
15. Jones, M.; Winne, P.H. (1992) Adaptive learning environments: foundations and frontiers. F85
16. Lave, J., Wenger, E. (1991) Situated learning: legitimate peripheral participation. Cambridge, UK: Cambridge University Press
17. Leroux, P. (1992) A new development of control technology. In F116
18. Leroux, P. (1992) Cooperation between pupil and expert system to drive a micro-robot. Proc. 1st International Conference on Technology Education, Weimar, 25-30 April
19. Leroux, P., Bruneau J. (1992) Activating a micro-robot without using a programming language. Proc. ICCAL 92, Wolfville, 17-20 June
20. Major, N.P. (1993) Reconstructing teaching strategies with COCA. Proc. AIED 93. University of Edinburgh
21. Major, N.P. (1993) Teachers and intelligent tutoring systems. Proc. PEG 93. Heriot-Watt University, Edinburgh

22. Major, N.P., Reichgelt, H. (1992) COCA - a shell for intelligent tutoring systems. In: Frasson, C., Gauthier, G., McCalla, G.I. (eds.) Proc. ITS 92, Lecture Notes in Computer Science, Vol. 608. Berlin: Springer Verlag

23. Nicaud, J.F., Vivet M. (1988) Les tuteurs intelligents: réalisations et tendances de recherche, synthèse sur les tuteurs intelligents. Revue Techniques et Science Informatiques 7(1), January

24. O'Shea, T., Bornat, R., Du Boulay, B., Eisenstadt, M., Page I. (1984) Tools for creating intelligent computer tutors. In: Elithor, Banerjii (eds.) Human and artificial intelligence. North-Holland

25. Papert, S. (1980) Mindstorms: children, computers and powerful ideas. New York: Basic Books

26. Ponta, D., Parodi G.C., Ponta M. (1991) A learner interface for an electronic simulation environment. Proc. IFIP TC3, Alesund, July

27. Self, J. (1986) Artificial intelligence, its potential in education and training. Proc. 5ième symposium canadien sur la technologie pédagogique, Ottawa, 5-7 May, pp. 69-77

28. Self, J. (1988) Bypassing the intractable problem of student modelling. Proc. ITS 88, Montreal, June, pp. 18-24

29. Self, J.(1988) Student models: what use are they? In Artificial intelligence tools in education, Proc. IFIP-TC3, Frascati, May 1987. North-Holland

30. Self, J.(1989) The case of formalising student models (and ITS generally). Proc. Int. Conf. AI and Education, Amsterdam, 24-26 May, p. 244

31. Sleeman, D., Brown J.S. (1982) Intelligent tutoring systems. London: Academic Press

32. Teutsch, P. (1993) Interaction issues in computer assisted language learning systems. Proc. ICCE'93, Taiwan, December

33. Verdejo, M.F. (1992) A framework for instructional planning and discourse modelling . In F91, pp.146-170

34. Vivet, M. (1988) Knowledge based tutors: towards the design of a shell... In H. Mandl (ed.) special issue of International Journal for Educational Research and Instruction 12(8), 839-850

35. Vivet, M. (1988) Reasoned explanations need reasoning on reasoning and reasoning on the student. In: Artificial intelligence tools in education. Workshop IFIP-TC3, Frascati, May 1987. North-Holland

36. Vivet, M. (1989) Which goal, which pedagogical attitudes with micro-robots in a class room? In F86

37. Vivet, M. (1990) Learning with micro-robotics activities. In F

38. Vivet, M. (1990) Uses of ITS: Which role for the teacher? In F91, pp. 171-182

39. Vivet, M. (1991) Learning science and engineering with open knowledge based systems. In: E. Forte (ed.) Proc. CALISCE '91, Lausanne, 9-11 September (LEAO/EPFL). Presses polytechniques et universitaires Romandes, pp. 53-64

40. Vivet, M. (1991) Knowledge based systems for education: taking in account the learner's context. Proc. PEG-91, Rappallo, May. Special issue of Computers & Education

41. Vivet, M. (1992) Examination of two ways for research in advanced educational technology. In F96, pp. 177-190

42. Vivet, M., Leroux P. (1993) Relations in the triangle learner-teacher-microrobots based learning environment. Proc. 10th International Conference on Technology and Education, MIT, 21-24 March
43. Vivet, M., Leroux P., Hubert O., Morandeau J., Parmentier C. (1993) Teleassistance of trainees in an SME: a case study. Proc. Teleteaching '93, IFIP-TC3, 3rd International Conference and Exhibition in Learning and Working Independent of Time and Distance, Trondheim, Norway, 20-25 August
44. Vivet, M. (1992) Educational uses of control technology. In F116
45. Vivet, M., Parmentier, C. (1991) Low qualified adults in computer integrated enterprise: an example of in-service training. IFIP TC3/WG3.4, Alesund, Norway, 1-5 July 1991. In: B.Z. Barta, H. Haugen (eds.) Training: from computer aided design to computer integrated enterprise, pp. 261-272. North-Holland
46. Wenger, E. (1987) Artificial intelligence and tutoring systems: Computational and cognitive approaches to the communication of knowledge. Morgan Kaufmann
47. Winans, R.T., Whitaker, E.T., Bonnel, R.D. (1988) Theories of learning in computer-aided instruction. Proc. 5th International Conference on Technology and Education, Edinburgh, March 1988, pp. 86-89
48. Yazdani, M. (1986) Intelligent tutoring systems survey. AI Review 1, 43-52

12

Implementing Instructional Models in Computer-Based Learning Environments: A Case Study in Problem Selection

Jeroen J. G. van Merriënboer, Jaap Jan Luursema

University of Twente, Department of Instructional Technology
P.O. Box 217, 7500 AE Enschede, The Netherlands

Abstract: This chapter is concerned with instructional models, that is, implemented process models of instruction that offer some articulation of the didactic expertise involved. Three basic problems in the implementation of instructional models are discussed. Possible solutions to those problems are illustrated by a description of CASCO, a computer-based learning environment for introductory computer programming. Our focus is on the application of fuzzy set theory and fuzzy logic in the implementation of so-called Fuzzy Logic Instructional Models (FLIM's). To illustrate this approach, an in-depth discussion is provided of CASCO's Problem Selection model.

Keywords: Instructional models, didactic expertise, intelligent tutoring systems (ITS), fuzzy set theory, fuzzy logic, intelligent task generation, problem selection, artificial intelligence (AI), training strategies, introductory computer programming

1 Introduction

In the field of Advanced Educational Technology, there is a growing interest in articulated instructional models for computer-based learning environments (Dijkstra, Krammer, & Van Merriënboer, 1992; Van Merriënboer, 1994). Such models should offer an explicit representation of knowledge concerning one or more aspects of instruction. Their main advantage lies in the fact that they make it possible to build computer-based learning environments with a high flexibility to implement various instructional strategies, that is, they offer the opportunity to easily modify applied strategies and to perform systematic research on their effectiveness.

But unfortunately, the formulation of instructional models has proved to be a hard problem. Most Intelligent Tutoring Systems (ITS) or related computer-based instructional systems do embody a well-defined set of instructional principles, but those principles are wired in the system rather than *interpreted* by it. As a result, changing the applied instructional strategies often requires rebuilding most of the system. In our opinion, at least three issues explain the lack of implemented process models of instruction that offer some articulation of the expertise involved. These are the following:

Issue 1: Integrated Objectives. Intelligent Tutoring Systems often aim at the teaching of complex cognitive skills (e.g., computer programming, mathematical problem solving, fault management, etc.), which involve several constituent skills and are thus characterized by highly integrated sets of different instructional goals. With regard to training strategies, such complex cognitive skills typically ask for a mix between plan-based and opportunistic approaches. While plan-based approaches carefully manipulate the sequence of experiences, problems, or cases through which the learner is expected to acquire expertise, opportunistic approaches mainly take advantage of the teaching opportunities that arise in the context of some activity in which the student is engaged (Wenger, 1987, pp. 398-399). While most ITS's follow some version of an opportunistic approach, there have also been several successful attempts to use a plan-based approach (e.g., BIP; Barr, Beard, & Atkinson, 1976, see also Halff, 1988, MHO; Lesgold, Bonar, Ivill, & Bowen, 1987). However, it is currently far from clear how knowledge about instruction can best be represented in order to enable a synergy between plan-based and opportunistic approaches.

Issue 2: Data and Algorithms. While it is generally agreed upon that instructional models in computer-based learning environments should make decisions regarding instructional actions to be taken by the system, the border between the data, that is, the resources on which the decisions are based, and the algorithms is often difficult to indicate. For instance, in classical Computer Assisted Instruction (CAI) one might view the branching structure as a highly inarticulate instructional algorithm and the frames or screens as data. On the other hand, in Instructional Transaction Theory (Merrill, Li, & Jones, 1992) so-called instructional transaction shells may be viewed as algorithms that treat particular classes of subject matter content as their data. And furthermore, is should be clear that in adaptive instructional systems the knowledge state of the learner (as specified in a "student model") is also an important resource to base instructional actions on. To sum up, it is clear that information about instructional strategies and tactics, the structure of the domain to-be-taught, and the knowledge state of the learner are all important resources for an instructional model. But at the same time, it is debatable what must be considered as the data for those models and what must be considered as their algorithms.

Issue 3: Vague Knowledge. A last – but certainly not least – issue pertains to the fact that human teachers' expertise is an extremely complex skill of which little is understood. First, human decision making in instruction is usually based on incomplete or ambiguous information with regard to the knowledge state or motivational state of the learner (i.e., student model), the structure of the domain being taught, and other factors of interest; and second, the knowledge of instruction is often highly implicit or has the form of rough "rules-of-thumb". This situation leads some researchers to the conclusion that the state of knowledge about learning and instruction is too vague, implicit, and immature to represent this knowledge in instructional models (Merrill, Li, & Jones, 1990). But on the other hand, one might argue that computational models of instruction should try to justify the "fuzziness" of decision making in instruction.

The goal of this chapter is to introduce a representational formalism for instructional models that may overcome, at least to some degree, the issues mentioned above. The structure of our discourse is as follows. Section 2 provides a brief description of CASCO, a computer-based learning environment for introductory computer programming. Section 3 returns to the three issues and elaborates on the approaches that we have followed in the CASCO project to solve them; here, the focus will be on the notion of Fuzzy Logic Instructional Models (FLIM's) as an approach to model and articulate knowledge of instruction. A discussion of our approach and some suggestions for future research are included in Section 4.

2 An Outline of CASCO

CASCO is a learning environment for introductory computer programming (Van Merriënboer, Krammer, & Maaswinkel, 1993; Van Merriënboer, Luursema, Kingma, Houweling, & De Vries, 1994). CASCO's basic approach is to select programming problems and to dynamically generate programming assignments that are tailored to the particular needs of an individual learner; in addition, the learner is supported while working on the assignments. Whereas CASCO's architecture allows for the implementation of different training strategies, it is currently set up to employ the Completion Strategy (Van Merriënboer & Krammer, 1990; Van Merriënboer & Paas, 1990). According to this strategy, learners have to complete or extend increasingly larger parts of well-designed, well-readable, but incomplete computer programs; the assignments also serve to illustrate and explain new programming concepts and to ask questions to the learner about the programming language or programming task. In several experiments, the Completion Strategy proved to be quite effective (e.g., Van Merriënboer, 1990; Van Merriënboer & De Croock, 1992).

Figure 1 provides a schematic overview of CASCO. The three main databases in CASCO are a problem database, which contains the problem solving products (e.g., problems, solutions) that might be presented to the learner, a content database, which contains the to-be-learned elements to convey to the learner (called *learning elements*), and a delivery database, which contains the delivery templates

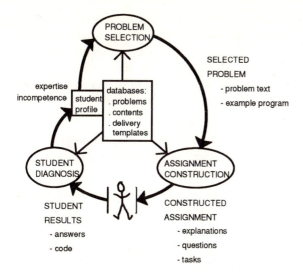

Figure 1. CASCO's global architecture, indicating its main databases and processes

that may be used to communicate the problem solving products and the learning elements. In addition, it may be seen from Fig. 1 that CASCO cycles through three main processes: Student diagnosis, in which the learner's results on the last assignment are used to update a student profile; problem selection, in which a new problem is selected from the problem database, and assignment construction, in which the selected problem is used to generate a new programming assignment to present to the learner. In this section we briefly discuss, in order, the training strategy currently applied by CASCO, its main databases, and its main processes.

2.1 Training Strategy Currently Applied in CASCO

In the Completion Strategy, so-called *completion assignments* are the basic building blocks of instruction[1]. They may consist of the following five elements:

1. *Problem description.* This is a description of a programming problem in natural language. This is the only element that must be present in each assignment. Note that a clear distinction is made between the concepts "problem" and "assignment": The term assignment is used in a much broader sense because it also pertains to the following four elements.

---

[1]Note that CASCO is a loose acronym for Completion ASsignment COnstructor; in Dutch, it also denotes to a "body" or "frame" that must be further specified, indicating the possibility to implement different training strategies within CASCO.

2. *Program code.* This is an example solution to the posed problem in the form of a computer program. The program code may be either complete (i.e., a well-designed, well-readable and executable computer program), incomplete (a partial program that must be completed by the learner), or absent (in which case the learner has to design and implement the whole program). In general, the program code that is provided to the learner becomes more and more incomplete as the learner's programming experience increases.

3. *Explanations.* New features of the programming language or the programming task may be explained. According to the Completion Strategy, explanations are always illustrated by – parts of – the example program.

4. *Questions.* Either multiple-choice questions or open questions may be asked on the working and the structure of – parts of – the example program.

5. *Instructional Tasks.* Tasks may be given to the learner to run, test, change or extend the program code (if it is complete), to complete – parts of – the program code (if it is incomplete), or to design and code a program (if the example program is absent).

Obviously, this basic scheme allows for a great diversity in completion assignments. At one extreme, an assignment may take the form of a fully worked-out example if it consists of a problem description, complete program code, and one or more explanations on new programming concepts that are illustrated in the program code. At the other extreme, an assignment may take the form of a conventional programming problem if it only consists of a problem description and the task to design and code a program. But as a rule, completion assignments contain a problem description, incomplete program code, and a – small – number of explanations, questions, and tasks.

2.2 CASCO's Main Databases

Problem Database. Essentially, the problem database is simply a set of problems that might be presented to the learner. For each problem, four kinds of problem solving products may be distinguished: (1) a problem description in natural language (which is seen as a zero-order product), (2) a goal decomposition, (3) a specification of the solution in terms of programming plans, and (4) the – executable – program code that offers a correct solution to the posed problem (Krammer, Van Merriënboer, & Maaswinkel, 1994).

In the Completion Strategy, only the first (problem description) and the last (program code) problem solving products are used and thus specified in the problem database. In addition, each problem is characterized by its name or title and a list of learning elements that must be applied by the learner to solve this particular problem and reach a correct solution. Each of these learning elements is described in the content database. In the current version of CASCO, the problem database contains 50 problems. New problems which use learning elements that are already specified in the content database can easily be added; if new learning elements are used by these problems, those learning elements must be added to the content database.

Content Database. The content database contains the learning elements that must be communicated to the learner; it is assumed to model the knowledge concerning the programming language and the programming task. In CASCO, seven different kinds of learning elements may be distinguished (see Krammer, Van Merriënboer, & Maaswinkel, 1994). In the so-called subject matter category, these are (1) programming goals, (2) programming plans, and (3) syntax rules; in the cognitive strategy category, these are (4) analysis heuristics, (5) plan principles, (6) discourse rules, and (7) test heuristics.

In the Completion Strategy, only three of these learning elements are used and thus specified in the content database: programming plans, syntax rules (including elementary commands of the programming language), and discourse rules. Programming plans are schematic descriptions of the structure of a particular piece of program code to accomplishes a specific goal of the program (Soloway, 1985); syntax rules refer to the syntax of elementary language commands, and discourse rules are rules-of-thumb which prescribe how to make a program easily readable and comprehensible.

All learning elements that are used by the problems as specified in the problem database are described in the content database. Five types of relations are defined between the learning elements in the content database:

1. *uses* – indicates that another learning element is a prerequisite for this learning element;
2. *together-with* – indicates that another learning element must preferably be presented together with this learning element;
3. *not-with* – indicates that another learning element must preferably not be presented together with this learning element;
4. *resembles* – indicates that another learning element looks like this learning element, and
5. *preferably-after* – indicates that another learning element should preferably be presented before this learning element; this relation is somewhat like the *uses* relation but does not pose a "hard" prerequisite.

Furthermore, each learning element is characterized by its name and title, a formal parameter list (in order to relate the learning element to a particular problem; actual parameters are provided in the problem database), and a list of delivery templates which may be used to communicate the content to the learner. Each of these delivery templates is described in the delivery database.

Delivery Database. This database contains the delivery templates that are used to communicate problem solving products (problem descriptions, program code) and learning elements (programming plans, syntax rules, discourse rules) to the learner. In the Completion Strategy, delivery templates concern problem specifications, example programs, explanations, questions, and instructional tasks. Levels may be specified for those templates. No levels are used for the problem specification, where the template is nothing more than an empty window that may be filled with the selected problem description, and the example program, where the template is actually the editor in which the program code is presented to the learner. Two levels are used for explanation templates (normal, remediation), three levels for question templates (simple, normal, difficult), and four levels for task templates (narrow context, normal context, broad context, no context).

In a simple sense, CASCO's main task is to instantiate particular delivery templates with particular problem solving products or learning elements in order to generate instruction, yielding, for instance, a particular problem description, a normal explanation on a discourse rule that is used for the first time, a difficult question on a programming plan that has been used before, an instructional task to write a piece of code in a situation where little context (i.e., surrounding code) is offered, and so forth. The instantiated delivery templates are also called *deliveries*.

As a requirement of the Completion Strategy, as well as some other training strategies that might be implemented in CASCO, explanations, questions, and tasks are coupled to the program code that is presented to the learner. While specifying the templates in the delivery database, three options are available to reach this goal. First, a part of the presented program code that is related to the delivery may be highlighted. Second, parameters may be used in the texts accompanying deliveries in order to refer to particular elements of the presented program code (line numbers, variable names, names of procedures, etc.). And third, parts of the presented program code may be changed, or simply deleted. The last option is mainly used in task templates: It enables the deletion of the program code that must subsequently be completed by the learner.

2.3 CASCO's Main Processes

Student diagnosis and student profile. While the learner is working on an assignment, student diagnosis takes place in order to update the student profile. For the Completion Strategy, the learner's results on questions and instructional tasks form the input of student diagnosis. Essentially, the student profile is an overlay of the content database. For all learning elements that have already been presented to the learner, the so-called Expertise and Incompetence are computed, where Expertise indicates the learner's proficiency in correctly using a particular learning element while Incompetence indicates the learner's tendency to make errors with a particular learning element.

For each learning element, the Expertise and Incompetence are further modelled as *fuzzy sets*; the truth value of the membership of those sets may range between 0 and 1. In Fig. 2, the fuzzy sets and membership values for Expertise (with the sets Prenovice, Beginner, Intermediate, and Expert) and Incompetence (with the sets Understander, Go-between, Fault-maker, and Incompetent) are given. As may

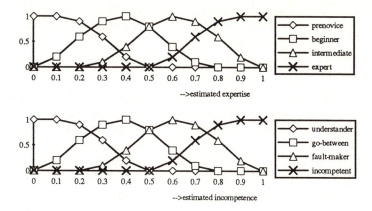

Figure 2. Membership values of the fuzzy sets derived from the Expertise and Incompetence on a particular learning element

be seen from this figure, membership in multiple, or even contradictory sets is possible. For instance, a learner with an Expertise of .4 for a particular learning element is categorized as a Prenovice, an Intermediate, as well as a Beginner – but as a Beginner in the highest degree. As a result of student diagnosis, the student profile characterizes each learning element either as a new learning element that has not yet been presented to the learner (then, it is a member of the set IsNew), or as a presented learning element, with associated membership values for the fuzzy sets within Expertise and Incompetence.

Problem Selection. The goal of the selection process is to pick from the problem database one or more problems that are suitable to present to the learner. In order to determine if a problem is suitable, both characteristics of the learner and characteristics of the problem, in relation to the structure of the content database, are of importance. With regard to the learner, important aspects are assumed to be the learning elements the learner is already practicing but does not master yet (i.e., the Learning Set) and the learning elements the learner does not master but makes mistakes with (i.e., the Incompetence Set). As a rule of thumb, a new problem should not contain too many new learning elements if the Learning Set is already large, and it should contain no new learning elements but focus on the learning elements that are not well understood if the Incompetence Set is not empty.

With regard to the problems themselves, important aspects are assumed to be their so-called "presentability" and "practicability". For example, the presentability of a problem is high if it contains new learning elements for which the learner has the prerequisite knowledge available; on the other hand, it is low if it uses learning elements that have a *uses* relation to other new learning elements or a *not--with* relation with other learning elements in the same problem. As another example, the practicability of a problem is high if it contains learning elements that are known by the learner but in which the learner is not yet an expert; on the other hand, it is low if the problem neither contains known learning elements nor elements in which the learner is already assumed to be an expert. As a rule of thumb, a new problem should combine a high presentability with a high practicability.

Assignment Construction. After a problem has been selected from the problem database, CASCO constructs an assignment by instantiating appropriate delivery templates with problem solving products or learning elements. As may be seen from Fig. 3, CASCO presents separate windows to the learner for the problem description and, if applicable, the – partial – program code and each of the explanations, questions, and tasks. The learner receives the deliveries that have to be studied (in case of an explanation) or performed (in case of a question or task) in an ordered sequence by clicking on the "Continue" option in the main menu; however, the learner may also use the "Window" submenu to select deliveries in an arbitrary order.

The window containing the program code is a full-fledged editor, enabling the learner to behave in exactly the same way as in a normal programming environment; the program code can be modified without any restrictions. At all times, the learner is able to run and test the program by selecting the option "Try your program" from the "Assignment" submenu. The learner is also able to go to the

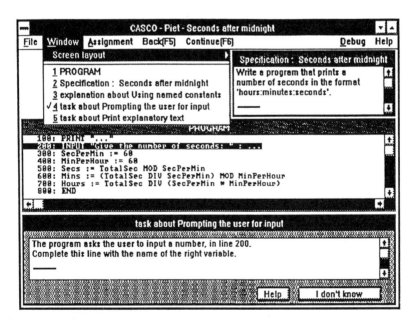

Figure 3. CASCO's presentation of (part of) a Completion Assignment, containing a problem description, a to-be-completed program, and one instructional task

next assignment whenever desired: The learner receives a warning if this option is chosen before questions are answered and/or tasks are performed. If the warning is neglected, negative results are diagnosed which has effects on the selection and construction of the next assignment.

3   Three Issues Revisited

In the previous section, we provided a global description of CASCO's functional architecture, its main databases and processes, and its interface to the learner. The present section will discuss how CASCO is dealing with the three issues presented in the introduction, namely, the issue of integrated objectives, the issue of data and algorithms, and the issue of vague knowledge. Here, our focus will be on the articulation of didactic expertise in CASCO, that is, its instructional models.

3.1 Dealing with the Issue of Integrated Objectives

Our basic answer to the issue of integrated objectives is *decomposition*. The design of CASCO is inspired by the Four-Component Instructional Design model (4C/ID-Model; Van Merriënboer, Jelsma, & Paas, 1992). This model focuses on the analysis of a to-be-trained complex cognitive skill (for CASCO, computer programming) in a process of principled skill decomposition and the conversion into a training strategy which is build up from a number of instructional tactics and ready for implementation.

The 4C/ID-model distinguishes categories of instructional tactics and allows for the implementation of those tactics as separate instructional models, pertaining to different categories of delivery templates. As may be seen from Fig. 4, CASCO currently uses distinct instructional models for (1) the selection of problems, the construction of (2) instructional tasks, (3) questions, and (4) explanations, and the generation of feedback on (5) answered questions and (6) finished tasks. Each model is easy to modify and other models may simply be added. The main role of the Model Manager is to solve conflicts that may occur if distinct instructional models yield incompatible instructional actions.

This approach is apt to implement training strategies that mix opportunistic and plan-based approaches. As an example, one might consider a case where the Selection Model is strictly plan-based, by ordering the problems that the learner receives on the basis of the structure of the domain (e.g., prerequisites first, simple first, important first) without taking the knowledge state of the student into account; but at the same time the feedback model may be highly opportunistic, for instance by directly providing informative feedback when the knowledge state of the learner indicates a particular misconception. In addition, plan-based and opportunistic approaches may be combined in the same model. The Selection

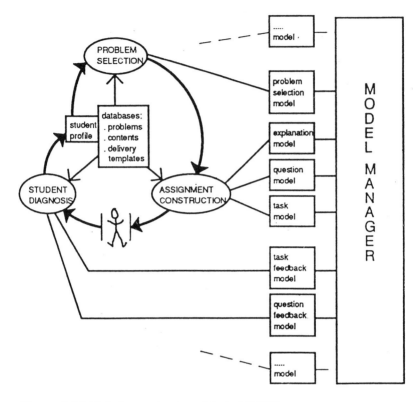

Figure 4. Multiple instructional models in CASCO

Model, for example, becomes more opportunistic if the selection of a next problem is driven by the assumed knowledge state of the student.

3.2 Dealing with the Issue of Data and Algorithms

Actually, this issue touches the kernel of what an instructional model is – and what it should do. In our approach, the main function of an instructional model is to consider the *appropriateness* to instantiate particular delivery templates with particular content elements (i.e., either learning elements or problem solving products) and to decide which deliveries will be presented to the learner. CASCO supports this approach by making a clear distinction between content elements (specified in the problem database and the content database) and delivery templates (specified in the delivery database), and by enabling the use of distinct instructional models for different categories of delivery templates.

In addition, it should be clear that the main resources for the model's reasoning concern the assumed knowledge state of the learner (student profile) and the structure of the domain (problem and content database). This leads us to our definition of instructional models: Computational process models of instruction that consider the appropriateness to present particular deliveries to the learner (i.e., instantiate particular delivery templates with particular content elements), given the available information on the knowledge state of the learner and/or the structure of the domain to-be-taught. But as mentioned before, it is exactly the information on the knowledge state of the learner and the structure of the domain that is often vague and incomplete. Our approach is to use fuzzy sets to denote this information, enabling the use of natural language-like concepts to describe it. This will be further explained in the next section.

3.3 Dealing with the Issue of Vague Knowledge

In order to deal with the issue of vague knowledge, the instructional models in CASCO use fuzzy sets (as introduced by Zadeh, 1965) and fuzzy logic (for an introduction, see Brubaker, 1991; Zadeh & Kacprzyk, 1992). For this reason, we call them Fuzzy Logic Instructional Models (FLIM's). In the specification of FLIM's, three phases can be distinguished. First, the information for the model, pertaining to the knowledge state of the learner and the structure of the domain, is defined by the use of fuzzy sets. This process is also known as "fuzzification". Second, the rules must be formulated that specify the behavior of the model; these rules govern the "fuzzy reasoning" process regarding the appropriateness of particular instructional actions. And third, the model's outputs must be transformed to a usable format; this process is known as "defuzzification".

FLIM's derive much of their power from their ability to draw conclusions and generate responses based on vague or incomplete information regarding the student or the domain to-be-taught. Moreover, the behavior of a FLIM is represented in a simple and natural way, allowing the user of the system easy modification of the instructional strategies that are applied. In the remainder of this section, we will provide an in-depth description of CASCO's Selection Model to illustrate the nature of a FLIM (for a description of the other instructional models in CASCO, see Van Merriënboer, Luursema, Kingma, Houweling, & de Vries, in press). In order, a description will be provided of the fuzzification of information regarding the learner, the fuzzification of information regarding the domain, the rules that govern the fuzzy reasoning process, and the defuzzification of the model's output.

Information about the learner. As argued before, important aspects with regard to the learner are the number of learning elements the learner is already practicing but does not master yet (the Learning Set; if this is large, a new problem should not contain too many new learning elements) and the number of learning elements the learner makes mistakes with (the Incompetence Set; if this is large, a new problem should not contain new elements but focus on remediation of the ill-understood elements). Here, it should be recalled that a number of fuzzy sets is specified in the student profile. Amongst others, the following sets are known:

- Expert – members of this set are learning elements in which the learner is assumed to be an expert.
- Understander – members of this set are learning elements that the learner seems to understand.
- IsNew – members of this set are learning elements that have not yet been presented to the learner.

A particular learning element is a member of the Learning Set if it is neither a member of the set IsNew nor a member of the set Expert. The size of the Learning Set may then be computed as the *cardinality* of this set, which is simply defined as the sum of the membership degrees of all learning elements in the set:

$$\#LearningSet \quad = \quad \sum_{LeProfile} (\text{not L.IsNew and not L.Expert}) \tag{i1}$$

The same holds for the Incompetence Set. A particular learning element is a member of this set if it is neither a member of the set IsNew nor a member of the set Understander. The cardinality is then defined as:

$$\#IncompetenceSet \quad = \quad \sum_{LeProfile} (\text{not L.IsNew and not L.Understander}) \tag{i2}$$

Obviously, particular learning elements may be a member of the Learning Set or the Incompetence Set in a higher or lower degree. Actually, the same learning element may be a member of both the Learning Set and the Incompetence Set. The formulas i1 and i2 yield non-integer values because they are the result of a summation of membership degrees[1].

Information about the problems and the domain. While the information about the learner is independent of the particular problem one might consider for deliverance, there is also relevant information that relates the characteristics of a particular problem to the structure of the domain; this information is unique for each considered problem. As argued before, important aspects with regard to the problems in the problem database are their so-called presentability (e.g., a problem containing *new* learning elements, for which the learner has the prerequisite knowledge available, is appropriate to be presented) and practicability (e.g., a problem containing previously presented learning elements, which the learner does not yet master, is appropriate to be practiced). In order to model the presentability of a problem, three fuzzy sets are defined:

- IsTooDifficult – members of this set are new learning elements with a *uses* relation to other new learning elements (then, they are a member of the set

---

[1] It may seem strange to sum the results on a logical expression. However, the expressions yield membership values in the interval 0..1 because *not*, *and*, *or* and *and_also* are implemented as fuzzy operators: *not* A as $1 - A$, A *and* B as min(A, B), A *or* B as max(A, B), and A *and_also* B as A * B.

TooDifficult), or with a *not-with* relation to other new learning elements that are used in the same problem (then, they are a member of the set Collides).

- IsLessGood – members of this set are new learning elements with a *preferably-after* relation to other new learning elements.
- IsPresentable – members of this set are new learning elements for which the prerequisite learning elements are known by the learner.

It should be recalled that the relations between learning elements are specified in the content database. In analogy to the fuzzification of learner information, the cardinalities of the fuzzy sets may be computed to yield (fuzzy) information on the presentability of each problem:

$$\#\text{IsTooDifficult} \quad = \sum_{LeProblem} (\text{L.IsNew and (L.TooDifficult or L.Collides)}) \quad (i3)$$

$$\#\text{IsLessGood} \quad = \sum_{LeProblem} (\text{L.IsLessGood}) \quad (i4)$$

$$\#\text{IsPresentable} \quad = \sum_{LeProblem} (\text{L.IsNew and not L.TooDifficult and not L.Collides}) \quad (i5)$$

The same approach may be followed in order to model the practicability of a problem. Again, three fuzzy sets are defined:

- NeedsFurtherPractice – members of this set are previously presented learning elements in which the learner is not assumed to be an expert (analogously to the definition of the Learning Set).
- IsKnown – members of this set are previously presented learning elements in which the learner is assumed to be an expert.
- PracticeIsUrgent – members of this set are previously presented learning elements which the learner does not seem to understand (analogously to the definition of the Incompetence Set).

The cardinalities of those fuzzy sets are also computed to yield information on the practicability of each problem:

$$\#\text{NeedsFurtherPractice} = \sum_{LeProblem} (\text{L.NeedsFurtherPractice}) \quad (i6)$$

$$\#\text{IsKnown} \quad = \sum_{LeProblem} (\text{L.IsKnown}) \quad (i7)$$

$$\#\text{PracticeIsUrgent} \quad = \sum_{LeProblem} (\text{L.PracticeIsUrgent}) \quad (i8)$$

Then, we have completed the fuzzification process by giving fuzzy descriptions of the necessary information on the learner and the domain; this information is of interest to the reasoning process that is performed by the Selection Model in order to consider the appropriateness of problems. Obviously, other fuzzy sets than the ones described above may be defined: It is up to the designer of the instructional model to decide on the information that is believed to be relevant to the specified instructional model. In the next section, a description will be provided of the rules of the Selection Model.

The Selection Rules. The didactic approach in the Selection Model that is currently implemented in CASCO is pretty straightforward. It starts from four desirable properties of a to-be-selected problem, assuming that a good problem:

1. is not too difficult and
2. has not been presented to the learner recently before and
3. is suitable to remediate learning elements the learner makes mistakes with, or
4. is suitable to present new learning elements and to practice known learning elements.

These basic ideas about the appropriateness of a to-be-selected problem are reflected in the first rule:

ProblemIsSuited   =   not TooDifficult and not RecentlyUsed and
                        (SuitedForRemediation or SuitedForPresAndPract)   (r0)

Due to the use of fuzzy logic, the rule assigns a fuzzy value between 0 and 1 to ProblemIsSuited; the higher the value, the more appropriate the problem is. Starting from this rule, and the four properties of a suitable problem that are reflected in it, the Selection Model is further specified. With regard to the first main property, claiming that a problem must not be too difficult, two situations are distinguished that can make a problem too difficult: (1) the problem may use learning elements for which the prerequisite knowledge is missing or (2) the problem may use learning elements that are less good because other learning elements should be presented first. These new properties are reflected in the following rule:

TooDifficult   =   ManyDifficultLelts or ManyLessGoodLelts   (r1)

In the previous section, the cardinalities #IsTooDifficult (i3) and #IsLessGood (i4) were defined. Obviously, there must be positive relationships between the property ManyDifficultLelts and the cardinality of the fuzzy set IsTooDifficult, and between the property ManyLessGoodLelts and the cardinality of the fuzzy set IsLessGood. Those relationships may simply be denoted by the following rules:

ManyDifficultLelts   = $IsLarge_1$ (#IsTooDifficult)   (r1.1)

ManyLessGoodLelts   = $IsLarge_2$ (#IsLessGood)   (r1.2)

The functions $IsLarge_1$ and $IsLarge_2$ can be defined in many different ways. For instance, Figs. 5a, b and c represent, in order, curves for $Islarge_1$ that allow almost no too difficult learning elements, a moderate amount of too difficult learning elements, and a fairly large amount of too difficult learning elements before a particular problem is actually considered to be too difficult.

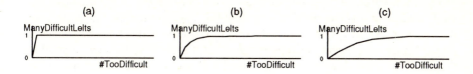

Figure 5. Some curves for IsLarge$_1$ that may be used to relate properties of problems to cardinalities of fuzzy sets (the function used is $1 - \exp(-c * \#\text{IsTooDifficult})$ with different values for c)

The same is true for the function IsLarge$_2$. However, the related curve should be less steep than for IsLarge$_1$ because too difficult learning elements will be typically considered to be more precarious than less good learning elements. This is shown in Figs. 6a and 6b. After defining the desired functions, the rules are ready to consider the appropriateness of a problem with regard to the first main property, namely, that it may not be too difficult (See Table 1 at the end of this section for an overview of the functions in CASCO's Selection Model and the used constants).

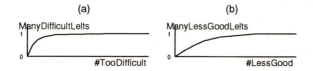

Figure 6. A comparison of possible curves for IsLarge$_1$ (a) and IsLarge$_2$ (b); the curve for IsLarge$_2$ should be flatter because less good learning elements are not so precarious as too difficult learning elements

This brings us to the second main property, namely, that an appropriate problem may not have been presented to the learner recently before. The rule defining the property RecentlyUsed is:

$$\text{RecentlyUsed} \quad = \quad \text{IsSmall}_1 \ (\text{nAgo}) \tag{r2}$$

Here, nAgo refers to the number of assignments that the same problem has been used before (note that this is simply an integer value not based on fuzzy sets - for instance, nAgo = 1 if the problem has been used for the foregoing assignment, nAgo = ∞ if the problem has never been used before). If the problem under consideration has never been used before, RecentlyUsed is 0; but if it has, the value of RecentlyUsed is related to nAgo by the function $IsSmall_1$. Of course, one might argue that it is bad practice to present the same problem twice: In this case, the function $IsSmall_1$ (nAgo) should simply yield the value 1 for all values of nAgo. On the other hand, one might argue that it is sensible to select a problem that has been used some time before; here, it should be remembered that in the Completion Strategy the same problem may be used to generate totally different assignments. Then, the function $IsSmall_1$ should be defined in such a way that the value of the property RecentlyUsed decreases as nAgo increases.

The third main property refers to the fact that, if a learner is assumed to need remediation, the selected problem must actually be suitable for remediation. In general, a problem is considered to be appropriate for remediation if it remediates many learning elements for which the learner makes mistakes and if it does so in a known context (i.e., no new learning elements are used in this problem; little other learning elements are further practiced). Those properties are reflected in the following rule:

$$\text{SuitedForRemediation} = \text{(RemediatesManyLelts and KnownContext)} \text{ and-also NeedsRemediation} \qquad (r3)$$

The property RemediatesManyLelts is positively related to the cardinality of the fuzzy set PracticeIsUrgent (i8). As for the properties ManyDifficultLelts and ManyLessGoodLelts, the relation may take different forms; but obviously, a problem with a higher cardinality of the fuzzy set PracticeIsUrgent enables the remediation of more learning elements:

$$\text{RemediatesManyLelts} = IsLarge_3 \text{ (#PracticeIsUrgent)} \qquad (r3.1)$$

Furthermore, it is assumed that a problem that is suited for remediation should do so in a known context. First, this means that the problem should not contain-many-new learning elements. To model this, the property FewNewLelts can simply be related to the number of new learning elements that are used in the problem (i.e., nNew, which is like nAgo an integer value not based on fuzzy sets); the relation should indicate that if the number of new learning elements in the problem under consideration increases, the value on the property FewNewLelts decreases. Second, the problem should not contain -many-learning elements that need further practice. To represent this in the model, the value of the property FewFurtherPractice can be related to the cardinality of the fuzzy set NeedsFurtherPractice (i6), again by an IsSmall function. These considerations are reflected in the following rules:

$$\text{KnownContext} = \text{FewNewLelts and FewFurtherPractice} \qquad (r3.2)$$

$$\text{FewNewLelts} = IsSmall_2 \text{ (nNew)} \qquad (r3.2.1)$$

$$\text{FewFurtherPractice} = IsSmall_3 \text{ (#NeedsFurtherPractice)} \qquad (r3.2.2)$$

The combined property (RemediatesManyLelts and KnownContext) as expressed in rule 3 indicates whether a problem can be used for remediation, but a problem is only suitable for remediation if the learner is also assumed to *need* remediation, that is, if the number of learning elements the learner makes mistakes with is relatively large. The learner's need for remediation may be seen as a function of the cardinality of the IncompetenceSet (i2), yielding the following rule:

$$\text{NeedsRemediation} \quad = \quad \text{IsLarge}_4 \text{ (\#IncompetenceSet)} \qquad \qquad \text{(r3.3)}$$

Here, the fact should be stressed that if a learner doesn't need remediation, a problem is not assumed to be suited for remediation - even if it uses each relevant ill-understood learning element in a fully known context. This is reflected by the expression "and-also NeedsRemediation" (see rule 3), where the fuzzy operator *and-also* is a "hard" version of a fuzzy *and*-operator that preserves information from the properties with the highest fuzzy values. For the normal fuzzy *and*-operator, A *and* B is implemented as min(A, B); A *and_also* B, on the other hand, is implemented as A * B. For example, suppose that two problems have fuzzy values on the combined property (RemediatesManyLelts and KnownContext) of, respectively, 1.0 and 0.75, but an identical value of .70 on the property NeedsRemediation. While the normal fuzzy *and*-operator should yield identical values of .70 on the property SuitedForRemediation for both problems, the fuzzy operator *and-also* yields values of .70 and .525 – correctly indicating that the first problem is more appropriate to select[1].

If remediation is not considered to be necessary, a regular assignment should be selected, that is, an assignment that is apt to present new learning elements or to practice learning elements that have been presented before. In addition, the selected problem should not lead to a Learning Set (i.e., number of not fully-mastered learning elements) that is too large. This fourth and last main property, called SuitedForPresAndPract, is expressed in the following rule:

$$\text{SuitedForPresAndPract} = \quad \text{(Presents or Practices and not New\#LearningSet} \\ \text{TooLarge)and-also not NeedsRemediation} \qquad \text{(r4)}$$

In contrast to the main property SuitedForRemediation (rule 3), the expression "and-also not NeedsRemediation" in this rule reflects the fact that the main property SuitedForPresAndPract can never receive a high value if the learner is assumed to need remediation. The properties Presents and Practices can, in order, be easily related to the cardinalities of the fuzzy sets IsPresentable (i5) and NeedsFurtherPractice (i6). They may be related to each other by IsLarge functions:

$$\text{Presents} \qquad = \quad \text{IsLarge}_5 \text{ (\#IsPresentable)} \qquad \qquad \text{(r4.1)}$$

$$\text{Practices} \qquad = \quad \text{IsLarge}_6 \text{ (\#NeedsFurtherPractice)} \qquad \qquad \text{(r4.2)}$$

---

[1] Thus, *and-also* has the following feature: If (A > B) and (C > 0), then (A *and-also* C) > (B *and-also* C).

Furthermore, a selected problem may not lead to a Learning Set that is too large, because it is assumed that working on too many learning elements that are not fully mastered may confuse the learner and hamper the learning process. This property is indicated in the following rule:

New#LearningSetTooLarge = IsTooLarge (New#LearningSet, Max)   (r4.3)

The function IsTooLarge in this rule is a new kind of function, indicating that a new learning set is considered to be too large if it amounts above a particular criterium, called Max. Currently, the maximum size of a new learning set is – somewhat arbitrarily – set at 6. The New#LearningSet may simply be defined as the sum of the cardinality of the existing Learning Set (i1) and the number of new learning elements used in the problem under consideration (i.e., nNew, as used in rule 3.2.1 before). This is reflected in the rule:

New#LearningSet   =   #LearningSet + nNew   (r4.3.1)

This completes the description of the rules that specify CASCO's Selection Model. To conclude this section, we would like to stress that it is not the rule base per se, but the chosen representational format that is, in our opinion, of main importance. The format allows for easy modification of the Selection Model on several levels. First, the model may be changed by changing the functions that are used to relate the model's rules to the information on the learner and the domain. The functions currently used in the Selection Model are shown in Table 1: these functions, and especially the constants used in them, are extremely simple to change. Second, the rule base itself may be altered: Because it consists of a relatively small set of understandable rules this is an easy task, permitting a relatively simple check on consistency. And finally, the Selection Model may be modified by the introduction of disparate information on the learner and/or the domain, requiring the definition of new fuzzy sets that are considered to be of interest[1]. In general, this is a more elaborate task because the potentiality to introduce new fuzzy sets is constrained by CASCO's model of the domain and its student profile.

Defuzzification. The Problem Selection Model as presented so far is able to consider for each problem in the problem database if it is an appropriate problem to deliver. Several heuristics are used to make the reasoning process more efficient in terms of computer time; however, a discussion of those heuristics falls beyond the scope of this article. The easiest way to defuzzify the model's output is to instantiate the problem specification window with the problem description associated with the most appropriate problem (i.e., the problem with the highest fuzzy value on the final property ProblemIsSuited). However, in order to secure a firm amount of learner control, CASCO presents a menu to the learner with the

---

[1] On the other hand, not all defined fuzzy sets need to be used by the model's rules. For instance, the fuzzy set IsKnown (i7) is not used in the specified rule base.

Table 1. Functions used in the selection model

Rule	Relation	Function	Where c is...
1.1	$IsLarge_1$(#IsTooDifficult)	$1 - \exp(c*\ \#IsTooDifficult)$	2.0
1.2	$IsLarge_2$(#IsLessGood)	$1 - \exp(c*\ \#IsLessGood)$	1.0
3.1	$IsLarge_3$(#IPracticeIsUrgent)	$1 - \exp(c*\ \#PracticeIsUrgent)$	2.0
3.3	$IsLarge_4$(IncompetenceSet)	$1 - \exp(c*\ \#IncopetenceSet)$	1.0
4.1	$IsLarge_5$(#IsPresentable)	$1 - \exp(c*\ \#IsPresentable)$	1.5
4.2	$IsLarge_6$(#NeedsFurtherPractice)	$1 - \exp(c*\#NeedsFurtherPractice)$	1.0
2	$IsSmall_1$(nAgo)	$\exp(-c*nAgo)$	0.4
3.2.1	$IsSmall_2$(nNew)	$\exp(-c*nNew)$	0.4
3.2.2	$IsSmall_3$(#NeedsFurtherPractice)	$\exp(-c*\#NeedsFurtherPractice)$	0.3
4.3	IsTooLarge(New#LearningSet, Max*)	$0.5 + \operatorname{atan}(c*(New\#LearningSet -6))/\pi$	0.3

*Max is set at 6.

titles of all problems that have a value on ProblemIsSuited that is not too far below the value of the problem found to be the most appropriate one (the border is usually set at 80%).

After the learner has selected a problem from the menu, the completion assignment that is related to the selected problem is constructed by some of CASCO's other instructional models (Explanation Model, Question Model, and Task Model). These models are represented in a format that is identical to that of the Selection Model (for a description, see Van Merriënboer, Luursema, Kingma, Houweling, & De Vries, in press). Here, it should be noted that some information on the selected problem (e.g., if it is mainly appropriate for presentation, practice, or remediation) is transferred to the Model Manager because it might affect the construction of the assignment.

4   Discussion and Future Directions

This chapter started from the idea that there is a need for explicit, articulated instructional models which enable an easy modification of the instructional principles that are applied in computer-based learning environments. However, there is also a lack of methods, techniques, and tools that may help in the design and implementation of such models. Three problems were discussed which may, at least partly, be responsible for this situation: The issue of integrated objectives, the issue of data and algorithms, and the issue of vague knowledge. Possible solutions to those problems were illustrated by a description of CASCO, a computer-based learning environment for introductory computer programming.

Our basic answer to the issue of integrated objectives is decomposition: separate instructional models are build which each support the performance and acquisition of constituent skills as identified in a task-analytical process of principled skill decomposition. The models pertain to different categories of delivery templates and are largely independent of each other. While this approach allows for a synergy between plan-based and opportunistic approaches to the design of instruction, is should be clear that the management of active instructional models is a far from trivial task. For instance, take a simple situation in which CASCO's Question Model considers it to be appropriate to pose a diagnostic question on a particular learning element, because there is some evidence that the learner has difficulties with its correct use. Simultaneously, the Explanation Model may consider it to be appropriate to provide some extra explanation on the same learning element; here, a conflict occurs between two instructional actions that must be solved by CASCO's Model Manager. In our current work, much effort is invested in the further development of this Model Manager, which may itself be conceived as a Fuzzy Logic Instructional Model that is operating on a higher didactic level.

Second, our answer to the issue of data and algorithms is quite straightforward. In the presented view, the main task of an instructional model is to consider the appropriateness to instantiate particular delivery templates with particular content elements (i.e., either problem solving products or learning elements), based on available, often incomplete "data" regarding the learner, the domain, and possibly other factors of interest. However, in our approach those data are not assumed to be directly given in a model of the domain or the learner. Instead, a highly significant task in the conceptualization and development of an instructional model pertains to the formulation of the *didactic concepts* that make up this data. In our opinion, much more empirical research is needed to investigate the important and meaningful concepts that teachers or instructional designers use in their reasoning about instruction.

Finally, our approach is characterized by the use of fuzzy sets to describe the didactic concepts that are used in an instructional model, so that natural language terms can be used to describe the relevant information. In addition, fuzzy logic is applied to model the reasoning about instruction so that the behavior of the model is specified by a relatively small set of meaningful rules, that are easily understandable by human beings. While fuzzy approaches have been applied in the field of student modelling (e.g., Gisolfi, Dattolo, & Balzano, 1992), we are not aware of other research that uses fuzzy approaches to model didactic expertise. Nonetheless, the results of the CASCO project support the basic idea that a fuzzy approach may lead to instructional models that are both articulated and relatively easy to implement, modify, and maintain.

As an important goal for future research and development, CASCO will be supplemented with a user-friendly interface that enables teachers or instructional designers to change applied instructional strategies by modification of the instructional models that make up those strategies. So, CASCO is envisaged to be a totally open system with interfaces to both learners and teachers, empowering the last to develop their own programming curricula and programming instruction. It is our firm conviction that tapping the expertise of experienced teachers or instructional designers, and offering them the tools to express this expertise, will ultimately lead to more effective computer-based learning environments.

Note

CASCO is an object-oriented program written in C++; it is running on a 486 PC under the Windows operating system. The programming language taught to the learners is UniComal (this is a kind of structured BASIC with Pascal-like control structures and data-types). CASCO and its documentation are freely available from the author. Please write to Jeroen van Merriënboer, University of Twente, Dept. of Instructional Technology, P.O. Box 217, 7500 AE Enschede, The Netherlands.

References

(NATO ASI Series F volumes are indicated by F and volume number)

Barr, A., Beard, M., & Atkinson, R.C. (1976) The computer as a tutorial laboratory: The Stanford BIP project. International Journal of Man-Machine Studies 8, 567-596

Brubaker, D.I. (1991) Introduction to fuzzy logic systems. Menlo Park, CA: The Huntington Group

Dijkstra, S., Krammer, H.P.M., & van Merriënboer, J.J.G. (eds.) (1992) Instructional models in computer-based learning environments. F104

Gisolfi, A., Dattolo, A., & Balzano, W. (1992) A fuzzy approach to student modelling. Computers in Education 19, 329-334

Halff, H.M. (1988) Curriculum and instruction in automated tutors. In: M C. Polson & J.J. Richardson (eds.) Foundations of intelligent tutoring systems, pp. 79-108. Hillsdale, NJ: Lawrence Erlbaum

Krammer, H.P.M., van Merriënboer, J.J.G., & Maaswinkel, R.M. (1994) Plan-based delivery composition in intelligent tutoring systems for introductory computer programming. In: J.J.G. van Merriënboer (ed.) Dutch research on knowledge-based instructional systems. Special issue, Computers in Human Behavior 10, 139-154

Lesgold, A.M., Bonar, J.G., Ivill, J.M., & Bowen, A. (1987) An intelligent tutoring system for electronics troubleshooting – DC circuit understanding. In: L.B. Resnick (ed.) Knowing and learning: Issues for the cognitive psychology of instruction. Hillsdale, NJ: Lawrence Erlbaum

Merrill, M.D., Li, Z., & Jones, M.K. (1990) Second generation instructional design (ID^2). Educational Technology 30(2), 26-31

Merrill, M.D., Li, Z., & Jones, M.K. (1992) An introduction to instructional transaction theory. In F104

Soloway, E. (1985) From problems to programs via plans: The content and structure of knowledge for introductory Lisp programming. Journal of Educational Computing Research 1, 157-172

Van Merriënboer, J.J.G. (1990) Strategies for programming instruction in high school: Program completion versus program generation. Journal of Educational Computing Research 6, 265-285

Van Merriënboer, J.J.G. (1994) (ed.) Dutch research on knowledge-based instructional systems. Special issue, Computers in Human Behavior 10

Van Merriënboer, J.J.G., & de Croock, M.B.M. (1992) Strategies for computer-based programming instruction: Program completion versus program generation. Journal of Educational Computing Research 8, 365-394

Van Merriënboer, J.J.G., Jelsma, O., & Paas, F.G.W.C. (1992) Training for reflective expertise: A four component instructional design model for complex cognitive skills. Educational Technology, Research & Development 40, 23-43

Van Merriënboer, J.J.G., & Krammer, H.P.M. (1990) The Completion Strategy in programming instruction: Theoretical and empirical support. In: S. Dijkstra, B.H.M. van Hout-Wolters, & P.C. van der Sijde (eds.) Research on instruction: Design and effects, pp. 45-61. Englewood Cliffs, NJ: Educational Technology Publications

Van Merriënboer, J.J.G., Krammer, H.P.M., & Maaswinkel, R M. (1993) Automating the planning and construction of programming assignments for teaching introductory computer programming. In F119

Van Merriënboer, J. J. G., Luursema, J. J., Kingma, H., Houweling, F., & de Vries, A.P. (1994) Fuzzy logic instructional models: The dynamic construction of programming assignments in CASCO. In F119

Van Merriënboer, J.J.G., & Paas, F.G.W.C. (1990) Automation and schema acquisition in learning elementary computer programming: Implications for the design of practice. Computers in Human Behavior 6, 273-289

Wenger, E. (1987) Artificial intelligence and tutoring systems. Los Altos, CA: Morgan Kaufmann

Zadeh, L.A. (1965) Fuzzy Sets. Information and Control 8, 338-353

Zadeh, L.A., & Kacprzyk, J. (1992) Fuzzy logic for the management of uncertainty. New York: John Wiley and Sons

13

Learning Interfaces

Philippe Duchastel

EDS, 31275 Heath Court, Beverly Hills, MI 48025, USA

Abstract: 'Learning interfaces' are the windows on the world through which a person views information and which cause a certain quality of learning to occur. Interfaces to learning are the cognitive artifacts, the resources for learning, that populate the learning environment and occasion learning. I believe this view of instruction and learning may prove useful in situating our efforts at improving learning within a larger perspective than we usually adopt in our professional dialogue.

Keywords: Interface, learning theory, instructional design, schooling

1 Introduction

While the notion of an interface may seem rather specific to those who deal directly with the design of computer-based learning environments, I propose that it is a concept that can be used in a very much wider context to convey the centrality of the person-environment interaction in the process of learning. Thus, I am not about to discourse on the process of learning a computer application interface, nor on the adaptability of interfaces through exposure to a user, but rather on how the central theme of interface design lies at the heart of advanced educational technology, and indeed goes far beyond it, too.

The future of educational technology is not to be found in evolving visions of the three technologies that give it shape, namely hardware, software, and process technologies, but rather in new ways of envisioning how these technologies can be used for the purpose of assisting learning. The focus on learning interface involves a slight but important paradigm shift that could be helpful in this respect. Paradigms are those ways of thinking that frame our conceptual explorations and lead us along certain paths rather than others.

Paradigm shifting, however, by its very nature, is always difficult. Indeed, initial reaction to this paper has been strongly polarized. One camp, which we might call the traditionalists (exemplified by O'Shea in his reaction statement)

considers the venture presented here to be unhelpful, if not downright misleading; the other camp, perhaps to be called the adventurists, values the potential benefit of radical rethinking and is willing to stretch the current envelope to see where such thinking might lead us (see for instance de Landsheere's reaction and the views of Derycke in this volume).

Whether exploring learning in terms of the concept of interface sounds reasonable or not will be for the reader to judge. What I deal with in this paper is old wine tasted from a new bottle. The decanting may prove hazardous, but the pleasure of the newly filtered vintage might just be appealing. The old wine I speak of is learning and the new bottle is the concept of interface.

An interface is generally though of as the surface level representation which a user interacts with in order to use a piece of equipment or a software application, with a view to engage in some purposeful task. An interface is, in this functional sense of the term, what lies between the tool's own function and the user herself. *The purpose of an interface is essentially to facilitate access to the tool's functionality, no matter whether we are dealing with physical tools or with mind tools.* Further, from a design perspective, an interface is itself a designed artifact, built a certain way within certain constraints, with a view to best accomplishing its mission.

Let me now generalize this common notion of interface. *An interface is the locus of interaction between a person and her environment.* Thus a telescope is an interface between a viewer and a scene; a steering wheel is an interface between a driver and a road; and a book is an interface between a reader and the ideas of an author. This is all over and beyond the fact that each of these artifacts has its own interface via the ergonomics of its specific design; thus, the artifact as seen here *is* the interface, although at a more abstract level of understanding. Going even further into abstraction, a play at the theater is the interface between the play-wright and the audience (over and beyond the physical interface of the actors, decor, and stage); likewise, the functions and responsibilities included in a job structure are the interface between an organization's aims and its actual accomplishments.

Now, let us stop a minute to consider how much we have just generalized the notion of interface. What we have done is to apply this notion to processes: viewing, driving, reading, entertaining, working. The core definitional element of an interface remains the same nonetheless: a design that facilitates access to a given functionality.

Thus, a steering wheel is an artifact that helps one follow the curves in the road, and a job is a design (an artifact, even if not a physical one) that will ensure accomplishment of some specific goals. Interfaces can be more or less concrete or abstract and more or less specific or global. All remain interfaces nevertheless.

2   Learning Interfaces

I am going now to jump straight into the theme of this paper, the specific area of learning interfaces. I will elaborate later on how learning as a process is an interface itself, that being an even more abstract perspective. For now though, let me start with concrete designs that facilitate the process of learning: what I have come over time to refer to as learning technologies. *Learning technologies are the designed environments that a person interacts with for learning purposes.*

Learning environments are not necessarily designed artifacts, but a great many of them are. Learning can occur at any time and under a large variety of circumstances. It is indeed a by-product of one's interaction with the surrounding world. For instance, just consider all the vocabulary a child picks up while playing with others, even before any formal schooling gets underway. This incidental learning is enormous in scope, but usually goes unattended. More concern is given to the formal curriculum, or formal training program, and hence to the design of learning opportunities to acquire formal knowledge.

With respect to the latter, I like to use the term 'learning *technology*' because it implies a disciplined and experienced way of creating those opportunities, via purposeful environments that support and guide learning. These designed learning vehicles can be more or less directive (that is, they can involve more or less learner-control of the interaction), an issue of very great importance in itself and which I have returned to many times in broadening my understanding of learning [8], but not the issue to pursue here.

In this light, a textbook is a learning technology, as is a drill & practice CAI program; the science lab is a learning technology, and so is an on-line hypermedia encyclopedia. All are designed artifacts that facilitate learning. Classroom instruction, by the way, is also usefully considered as a learning technology, and we will come to that later.

What I am getting at here is that even if learning technologies have been generally associated in the past with technological innovations like educational television, CAI, and so on, they are not limited to that. As a general category, they comprise any environment designed specifically for learning. And educational technology is the field of study whose object is learning technologies. You may have recognized an interesting phenomenon that is underway at this time: the term *educational technology* is seemingly making way to a new term, what is becoming known as *learning technology*, that term being used to refer to the same process technology as before [7,11].

The goal of learning technology is a design theory that is prescriptive for learning environments. Learning technology is considered here as the engineering discipline that is the repository for best methods and practices for the design of specific learning technologies. It is the accumulated '*How to*' that forms the basis for the professional side of the field.

Learning technologies are in evolution. Educational television has given way to CAI, which in turn has led to ICAI (Intelligent CAI), and parallel technology developments have led to the videodisk, to hypermedia, to hand-held computer access, and so on. As ever-new technological possibilities turn into concrete and economically accessible realities, instructional designers push the envelope of what is practically feasible in new learning designs. The field is thus continually confronted with change, sometimes mildly and sometimes radically. On the whole, that is an excellent situation to be in.

3   Status of Theory

Beyond the technical possibilities lie the common factors that organize the accumulated perspectives of the field and allow disciplined access to the practices that lead to success. This is the design theory specific to learning technologies, the *why, what,* and *how* of the field.

My assessment of theory is that we have not yet in hand anything that is very robust. We do have a number of theoretical viewpoints, such as those of Gagné and Briggs [10], Merrill [15], Reigeluth [19], all of which deal with global design issues affecting the building of learning technologies. They remain partial, however, and do not represent an integrated theory of learning technology. That theory remains elusive, but we may well see it emerge during this decade. I believe the setting is ripe for a bringing-together of the different strands at a high enough level of abstraction to create a somewhat full theory of learning technology. The nexus of such a theory may well reside, I believe, in what I am discussing here, namely the learning interface. Already for instance, we see it represented in Merrill's [15] focus on learning transactions, or in the evident interest, in the area of textual design, for adapting textual presentation to pedagogical purposes [12].

In the end, we return fully to psychology, the real mother science of instructional design, and even more so of learning technology. Indeed, the scientific basis of any design for learning is learning psychology. Remember B. F. Skinner's unbegotten attempts to build a learning technology on the foundation of his behaviorist learning theory. Well, his theory may have been way off the mark (by current thinking at any rate), but the logic of the process was indeed a good one. Gagné's attempt at instructional theory was similarly grounded in learning psychology, and indeed, all such attempts likewise are, even though they may not always be explicit in this regard.

The focus of learning psychology is the study of the constraints impinging on the interaction between learner and technology. We are dealing here with a very applied psychology, one that is ecologically valid and hence one that is far removed from the basic learning psychology generally found in the psychology journals. What is needed for our field is a learning theory that can account for the learner-technology interaction that is set up in any learning environment. That is a tall order, of course, given the variety of settings and diversity of factors that impinge on this kind of interaction.

Learning is an internal process that is stimulated and channeled by the external factors present. This view is basically that of Gagné [9], who considered learning in terms of its enabling conditions. From this perspective, learning can occur in any environment, from the most unstructured situation to the most formal learning environment. As mentioned earlier, a great deal of our learning happens all the time in informal settings as we continually interact with information that shapes our thinking, our specific knowledge, and our mental models. However, the quality of learning may well differ from situation to situation. Let me operationalize the notion of quality of learning in terms of appeal, intensity of effect, and time to learn. Designing learning environments should ideally optimize all three of these factors. It should lead to interesting interactions, to depth of understanding, and to succinct learning experiences.

Thus, the very reason we design learning environments is to create the conditions that are propitious to intensive learning. Through design, we provoke learning! Ideally, we design the technology so as to tap the best inner resources of the learner, so that what is activated is interest and meaningful learning, not patience and rote learning. Interest is a means to learning [14], while depth of understanding is the goal of education.

Learning psychology, at least the kind that is needed here, is itself not very robust. There are powerful theories of skill learning [1] and of meaningful learning [2,16], but only scant attention has been paid to how this all interacts with the epistemic curiosity side of the learner who is situated in a learning environment [3,17]. Fortunately, instructional design itself has taken up the challenge of this task, in the work for instance of Keller [13], or of Bransford and the Vanderbilt group [6], but the desired strong underpinning from learning psychology is still lacking. The practical, consequential issue boils down to this: What good is all the information or are all the activities present in a learning environment if the learner gets up and walks away from it, or otherwise disengages himself from the learning activity. This remains a major task for our field: to seek an integrated theory of applied learning that considers the full context of the learning environment.

4   Information and Learning

Information of course is the core of a learning environment, it is the abstract stuff that is juggled with and that eventually generates those internal mental models in the learner that we associate with conceptual learning. But consider how we authors and educators design that information by laying it out in certain ways rather than others, by emphasizing given elements rather than alternatives, and by representing it via particular media. That information we create is the core interface for learning!

Information as an interface is a mediator between knowledge and the learner.
Information is central to the communication of knowledge, as seen in the

following process cycle: Knowledge –> External Representation (i.e. *Information*) –> Learning Environment –> Learning (Internal Representation) –> Knowledge.

Information that is not simply data (for instance, the multiplication table) represents the interpretation of some set of complex knowledge by a particular author. No information of that kind is objective, but rather it is the constructed, external representation derived from one's own internal, construed understanding. Now, other than recognizing that there therefore exists an epistemology of didactics, as our colleague Balacheff well reminds us of, what is the importance of this perspective?

Well, for one, it reminds us of the Piagetian conception of learning where a constant dialectic between assimilation and accommodation results from the person's interaction with the external environment of information. All learning is internal juggling of perceptions stimulated by an external information environment. Knowledge is built up within the learner (or not, as the case may be) as a result of this juggling, and so personal knowledge is interpreted information.

That interpretation is what makes knowledge meaningful. It is also what makes it idiosyncratic to the learner, even if at times, it may well be similar to the knowledge expressed by others. The transfer of knowledge depicted above represents the voyage of knowledge from one individual to another through a pedagogical act of communication.

That view, though, is inappropriate. Communication is a mischievous analogy that in unproductive for theory. It holds too much of the notion of transvasing information, of pouring it from one mind to another, as if information were a commodity, even if an intellectual one. A learner cannot receive knowledge, only build it! So a better interpretation of the process is that one person, through exteriorizing her own knowledge, in fact creates an information environment that will serve as a potential interface for another person to personally build his own knowledge. The created information environment is a designed environment, and that is what I believe the epistemology of didactics is all about.

The very way in which the information is designed will certainly influence how it is interacted with. Here, we deal with two distinct, but ever-so related, facets of designed information. Learning psychology and learning environment design build heavily on the very relationship between these two facets, which are content selection and content representationi, the *what* and the *how* of information. If you think of it, the major recent theories of instructional design have dealt with this very issue: Gagné's hierarchies, Ausubel's advance organizers, Reigeluth's elaboration viewpoint all center on the structure of information presentation. The design of information is thus central to all.

5   Instruction and Learning Environments

The difference between informal learning in everyday life and more formal schooling and instruction lies in the oriented nature of the latter, its essential goal-directness or intentionality. All instruction is a motivated attempt on the part of one party, often globally conceived of as society acting through its agents, to influence the behavior and knowledge of another party, the student body. It is a deliberate and repeated attempt at influence, for the good of both parties. That is the mission of education, broadly speaking. *Teaching, or more broadly, instruction, involves organizing the learning interface (the interactions) with a view to having certain goals achieved.*

Both the instructional designer preparing the layout for an instructional computer-based simulation and the classroom teacher preparing an outline for the process of introducing certain concepts in class have foremost in mind what outcomes are sought in the learners. Both professionals design the learning environment to favor given outcomes, i.e. to encourage interactions that will lead to particular learning.

The whole idea of schooling lies in the third criterion of quality mentioned earlier for learning, to create on a certain scale a pattern of interactions that will enable succinct learning experiences. The mission of the school is to have a lot learned in a relatively short time frame. The core problem of schooling, perhaps, is that it may have overdone it: it may well have crafted instruction that is cognitively sound, in terms of information structure for instance, while having substantially overlooked the affective factors at play in the environment. Once again, we see the importance of the interest factor in learning, and more generally, of the crucial interrelationship between the elements of quality for learning [17].

Any learning environment must spark the epistemic curiosity of the learner if interest is to be sustained, so that the learner remains on task over time. This is the great challenge to educational technology today. The challenge is to create a framework and process for the design of learning environments that are inviting and have *pull* [4]. The technology that is evolving today is making this much more possible than it was formerly. That is the beauty of the age in which we live with respect to education and training.

My own interest in learning technologies stems from my belief in the power and importance of informal learning for the future. I believe that the student of 30 years from now will learn a great deal more than today from interactions with rich learning environments based in technology, rather than from the classroom learning environments that are still the norm currently. Teaching as we know it is in for a very rude awakening during the next few decades. This is not to say that teachers will disappear from schools, nor that schools become meaningless. What it does mean is that role transformations are about to happen in each case, with teachers and schools eventually doing what they do best (guiding and socializing the child), while leaving what they do poorly (instructing) to technology.

In industrial training circles, one of the usual steps in the process of developing a training plan is to sort out which tasks should be supported by job aids and which should instead be fully trained. The job aid approach has received a great deal of attention these past few years under the guise of what is called performance support systems. These are computer applications that facilitate job tasks by making available complex information to the user in an easily accessible manner. The availability of any job aid reduces the need for learning to some important extent. The job aid becomes one of the interfaces between the worker and the job; more accurately, it is part of the full interface between worker and job, one component among others. The main other one, of course, is the worker's own knowledge of the tasks to be performed. Training is thus only part of the total support for usage, involving internalization for the purpose of repeated usage.

Consider for a moment a rather odd question perhaps: Why do we learn? That is, why have we evolved as we have, increasing in both our learning capacity and in our need for learning? *Knowledge is an internalization of the external world for purposes of accessibility and ease of use.*

That rationale for learning becomes important in two circumstances: when the task is critical (e.g. handling an emergency procedure) or when the task is repeated often (many jobs fall into this pattern). Consider a simple illustration. Let's suppose I expect to go on a business trip to Korea in the coming year. I will learn a few words in Korean, but I will not invest much time in learning the language. Now suppose instead that my business calls me to travel to Korea one a month for the next 3 years; then, most likely, the effort to learn the language will seem reasonable. Expected usage of the knowledge is the determining factor in the learning decision. Knowledge is that internal asset that enables the doing once a decision to do something specific is made. *At a very general level, knowledge itself is an interface that is a mediator between the doer and the doing.*

6   Concluding Perspectives

Let us see now how we might tie all this together. Indeed, much of my discussion of interface has led us into high abstraction and while the philosophy of knowledge is interesting as such, what does it afford us in advancing our understanding of advanced educational technology? So, just let me recast the gist of the arguments I have made in this paper:

1. The context in which a learner interacts with information around her constitutes a learning environment, in fact an interface, which can be more or less propitious to learning.

2. Interfaces are designed artifacts, sometimes abstract cognitive ones, that constrain or direct the interaction between a person and that person's many environments, with the aim of facilitating the processes involved in a task.

3. Applied learning psychology studies this interaction between learner and environment and the resulting internal processes that constitute learning. As such, learning psychology guides learning technology in the design of learning environments.

4. Information is the exterior manifestation of knowledge, and as the core element of learning environments, it is central in structuring the processes that a learner engages in while building her own knowledge.

5. The field of learning technology is aimed at designing such environments to favor learning, either of the instructional kind that are highly goal-oriented, or of the informal kind that are unstructured and more open-ended.

6. Learning technologies, and more broadly all designed learning environments, are interfaces that structure the learner's interaction with information with the view of optimizing learning.

7. At a deeper level, the information designed into a learning environment is itself an abstract interface between a learner and the knowledge of others. How that information is designed is of the utmost importance, and indeed the focus of both learning psychology and learning technology.

8. A person's knowledge is itself an interface, at an abstract level, between that person and his or her actions; knowledge is an internal mediator to the world, juxtaposed next to all exterior interfaces.

Well, what does all this mean in practice? I believe there are implications for our conception of instruction, for our views of formal education, and for what we might see in the future.

The first involves an important shift in thinking from instruction being a process of communication, with its sender-receiver model, to learning being an interactive process through which mental models and skills are built and developed. Knowledge is not transferred, it can only be derived from an environment. This clearly centers the process on the learner, and in so doing, refocuses the entire field of educational technology. One of the strongest critique's of the field, and to my mind some of the best tonic we can use in this respect, is precisely aligned with this view. I am referring to Carroll's minimalist approach to instruction [5], in which personal meaning and interest take primacy over content coverage.

Now, there is nothing revolutionary about the view that we construct our knowledge actively and interactively. Only, our process technology (in fact, the field of instructional design) does not seem to have followed along with this evolving view. On the whole, and with all the danger inherent in generalization, it can be said that we still tend to design instruction within an older paradigm, despite our advanced rhetoric. Interface talk may help bring along the needed change.

The second implication lies in the area of schooling. As designed interfaces, learning environments are artifacts that have an embodiment of their own, with their own potential to affect learning. We can start characterizing different learning interfaces in terms of what they bring to this potential and how they may at times restrict it as well. For instance, the simulation interface brings realism and excitement to the interaction, whereas the traditional classroom brings social companionship and possibly competition for some. Each environment may emphasize certain features at the expense of others. It is just possible that the current schooling difficulties being experienced in many countries might be helped by rethinking the school as an interface, and not as a social structure unto itself.

Schooling is of course a complex interface composed of many interrelated facets operating at various levels. It nevertheless remains an interface, open to critical review and improvement just as any other [18].

The third implication considers the future of learning environments. As information artifacts, learning environments that are based in advanced technologies are becoming ever more accessible and powerful. In addition, as artifacts, they become improvable and thus grow in usefulness. Just like the knowledge base of an expert system is generally refined over time as many experts tinker with it, so too learning environments can be refined over time. This is why I essentially believe that future learning environments will be largely autonomous and freely accessible to learners without the socially-wrought constraints we see today. Learning interfaces then become key to success in education and training.

Perhaps the greatest benefit of considering learning and instruction, and even information and knowledge, as interfaces lies in leading us to refocus our attention as instructional designers. The need is to move away from the design and production of instructional products and focus instead on the design of interactions between learner and knowledge, in whatever form that knowledge is concretized and made available. Instructional design is classically oriented to the design of products, with incidental attention to the learning interaction. That orientation needs to be turned on its head: let us design and build interactions, with incidental attention to the products via which knowledge creation is stimulated. This involves a shift in the very object of design for our field.

Viewing our field as that of learning interfaces will hopefully also assist in gradually moving towards that integrated theory of applied learning that I indicated is so sorely lacking still today. It is certainly not the only view that will contribute to that redefinition of theory, but the hope is that it may spark interest in some radical rethinking yet to be undertaken. May the adventurists in advanced educational technology win out over the traditionalists!

References

(NATO ASI Series F volumes are indicated by F and volume number)

1. Anderson, J. (1983) The architecture of cognition. Cambridge, MA: Harvard University Press
2. Ausubel, D. (1968) Educational psychology: A cognitive view. New York: Holt, Rinehart and Winston
3. Berline, D. (1960) Conflict, arousal, and curiosity. New York: McGraw-Hill
4. Brown, J.S. (1983) Learning by doing revisited for electronic learning environments. In: M. White (ed.) The future of electronic learning. Hillsdale, NJ: Erlbaum
5. Carroll, J. (1990) The Nurnberg funnel. Cambridge, MA: MIT Press
6. CTGV (Cognition and Technology Group at Vanderbilt) (1993) Anchored instruction and situated cognition revisited. Educational Technology, March 1993, 52-68
7. Duchastel, P. (1989) The upcoming of learning technology. Canadian Journal of Educational Communication, Summer, 137-139
8. Duchastel, P. (1990) Assimilatory Tools for Informal Learning: Prospects in ICAI. Instructional Science 19, 3-9
9. Gagné, R. (1985) The conditions of learning. 4th ed. New York: Holt, Rinehart and Winston
10. Gagné, R., Briggs, L., and Wager, W. (1988) Principles of instructional design. 3rd ed. New York: Holt, Rinehart and Winston
11. Hannafin, M. (1992) Emerging technologies, ISD, and learning environments: Critical perspectives. Educational Technology Research and Development 40, 49-63
12. Jonassen, D. (ed.) (1982) The technology of text. Englewood Cliffs, NJ: Educational Technology Publications
13. Keller, J. (1983) Motivational design of instruction. In: C. Reigeluth (ed.) Instructional design theories and models: An overview of their current status. Hillsdale, NJ: Lawrence Erlbaum Associates
14. Malone, T. (1981) Toward a theory of intrinsically-motivating instruction. Cognitive Science 4, 333-369
15. Merrill, D., Li, Z., and Jones, M. (1991) Instructional transaction theory: An introduction. Educational Technology 31(6), 7-12
16. Novack, C. (1977) A theory of education. Ithaca, NY: Cornell University Press
17. Pintrich, P., Marx, R., and Boyle, R. (1993) Beyond cold conceptual change: The role of motivational beliefs and classroom conceptual factors in the process of conceptual change. Review of Educational Research, 63, 167-199
18. Reigeluth, C., Banathy, B., and Olson, J. (1993) Comprehensive systems design: A New Educational Technology. F95
19. Reigeluth, C., and Stein, F. (1983) The elaboration theory of instruction. In: C. Reigeluth (ed.) Instructional design theories and models: An overview of their current status. Hillsdale, NJ: Lawrence Erlbaum Associates

Subject Index

The NATO ASI Series F Special Programme on ADVANCED EDUCATIONAL TECHNOLOGY

NATO ASI Series F

Including Special Programmes on Sensory Systems for Robotic Control (ROB) and on Advanced Educational Technology (AET)

NATO ASI Series F

Including Special Programmes on Sensory Systems for Robotic Control (ROB) and on Advanced Educational Technology (AET)

NATO ASI Series F

Including Special Programmes on Sensory Systems for Robotic Control (ROB) and on Advanced Educational Technology (AET)

NATO ASI Series F